BLOOD WORK

ALSO BY HOLLY TUCKER

Pregnant Fictions:

Childbirth and the Fairy Tale in Early-Modern France

BLOOD WORK

A Tale of Medicine and Murder in the Scientific Revolution

HOLLY TUCKER

W. W. NORTON & COMPANY
New York · London

Printed in the United States of America
First Edition

For information about special discounts for bulk purchases, please contact
W. W. Norton Special Sales at specialsales@wwnorton.com or 800-233-4830

Manufacturing by Courier Westford
Book design by Chris Welch
Production manager: Julia Druskin

Library of Congress Cataloging-in-Publication Data

Tucker, Holly.
Blood work : a tale of medicine and murder in the
scientific revolution / Holly Tucker. — 1st ed.
p. ; cm.
Includes bibliographical references and index.
ISBN 978-0-393-07055-2 (hardcover)
1. Blood—Transfusion—Europe—History—17th century.
2. Denis, Jean Baptiste, d. 1704. I. Title.
[DNLM: 1. Denis, Jean Baptiste, d. 1704. 2. Blood Transfusion—history—Europe.
3. History, 17th Century—Europe. 4. Homicide—history—Europe.
5. Human Experimentation—history—Europe. 6. Public Opinion—Europe.
7. Science—history—Europe. WB 356]

RM171.T787 2011
615'.39—dc22

2010046340

W. W. Norton & Company, Inc.
500 Fifth Avenue, New York, N.Y. 10110
www.wwnorton.com

W. W. Norton & Company Ltd.
Castle House, 75/76 Wells Street, London W1T 3QT

1 2 3 4 5 6 7 8 9 0

For Audrey, heart and soul, moon and stars,

always and forever . . .

Very many maintain that all we know is still infinitely less than all that still remains unknown.

—William Harvey, *De motu cordis* (1628)

Blood is a juice of a very special kind.

—Johann Wolfgang von Goethe, *Faust, Part 1* (1808)

Contents

Note on Translations

In all circumstances, I have first relied on period translations of the original foreign-language texts cited. In the absence of a period translation, the remaining translations are my own. The source of the quote (printed translation or translated original) is indicated in the notes that accompany the text. Minor changes have been made to regularize spelling or typography, but these do not impact the original meaning.

Dramatis Personae

France

Louis de Basril (dates unknown)

outspoken lawyer in the Paris parliament

Jean-Baptiste Colbert (1619–83)

prime minister to Louis XIV

Jean-Baptiste Denis (c. 1635–1704)

physician and transfusionist

England

Robert Boyle (1627–91)

chemist; founding fellow of the Royal Society, corresponded with Richard Lower

Charles II (1630–85)

king of England; restored to the throne in 1661 after the execution of his father and the subsequent rule of Oliver Cromwell

Timothy Clarke (died 1672)

physician; founding fellow of the Royal Society, performed human infusion experiments with Christopher Wren

France

René Descartes (1596–1650)

philosopher; espoused the theory of mind-body dualism and the idea that the body was a machine

Paul Emmerez (died 1670)

surgeon; Denis' assistant

Nicolas Fouquet (1615–80)

Louis XIV's superintendent of finances; former political heir apparent to Prime Minister Mazarin

Pierre Gassendi (1592–1655)

philosopher; Descartes' intellectual rival, member of the Montmor Academy

Christian Huygens (1629–95)

astronomer, mathematician; former member of the Montmor Academy, founding member of Louis XIV's Academy of Sciences

England

Thomas Coxe (c. 1640–1730)

fellow of the Royal Society; replicated Lower's canine transfusions at the Royal Society with Edmund King

Arthur Coga (dates unknown)

first transfusion patient in England

William Harvey (1578–1657)

physician; announced discovery of blood circulation in 1628

Robert Hooke (1635–1703)

architect, microscopist; founding fellow of the Royal Society; former assistant to Thomas Willis, performed air-pump experiments on animals with Robert Boyle

Edmund King (1629–1709)

replicated Lower's canine transfusions at the Royal Society with Thomas Coxe

France

Guillaume Lamy (1644–83)

physician; member of the Paris Faculty of Medicine; outspoken critic of transfusion

Louis XIV (1638–1715)

king of France, also called the Sun King; began personal reign in 1661, following the death of Prime Minister Jules Mazarin

Henri-Martin de la Martinière (1634–76)

physician; once a doctor on corsair (pirate) ships; outspoken critic of transfusion

Antoine Mauroy (died 1668)

Denis' famous patient; died following a transfusion in April 1668

Perrine Mauroy (dates unknown)

wife of Antoine Mauroy

England

Richard Lower (1631–91)

physician; fellow of the Royal Society, performed first transfusion experiments in England

Henry Oldenburg (1619–77)

German-born diplomat and natural philosopher; secretary of the Royal Society

John Wilkins (1614–72)

founding member and first secretary of the Royal Society, with Henry Oldenburg

Thomas Willis (1621–75)

physician and anatomist; fellow of the Royal Society, studied the human brain with the help of Robert Hooke and Christopher Wren

Christopher Wren (1632–1723)

architect, astronomer, mathematician; founding fellow of the Royal Society; performed blood infusion experiments with Thomas Willis

France

Henri-Louis Habert de Montmor (c. 1600–79)

nobleman; founder of the Montmor Academy for the Sciences

René Moreau (1587–1656)

physician; member of the Paris Faculty of Medicine

Claude Perrault (1613–88)

physician, architect; member of the Paris Faculty of Medicine, founding member of Louis XIV's Academy of Science; performed canine transfusions in the king's library

Nicolas de la Reynie (1625–1704)

first police chief of Paris; appointed by Louis XIV

Samuel de Sorbière (1615–70)

permanent secretary of the Montmor Academy

Prologue

On December 14, 1799, America's first president awoke with a sore throat, which was soon accompanied by a fever. At six that morning, George Washington's doctors agreed it was time for a bloodletting. Eighteen ounces of blood later, the patient's condition had not improved, and he was bled twice more. Not long after, Washington was unable to breathe—medical historians believe that he suffered from an infection of the epiglottis—and a tracheotomy was performed. A fourth round of bloodletting followed, to no avail. Washington gasped for breath like a drowning man and died late that evening, around ten o'clock.[1]

Though we will never know whether Washington died of his illness or of the severe bloodletting he suffered during his "treatment," many historians would bet on the latter. His body was laid out in the family's formal parlor so that prominent visitors could pay their respects. Yet as the nation prepared to mourn its first president, others wondered if there was a way to bring him back to life.

When Washington's granddaughter, Mrs. Thomas Law, arrived the next morning, she brought with her a man who suggested the unthinkable. Dr. William Thornton, best known as architect of the U.S. Capitol, speculated that the president could be revived if both blood and air were returned to his corpse. Dr. Thornton suggested that Washington be warmed up "by degrees and by friction" so his blood might be coaxed to move once again through his body. Then Thornton proposed to "open a passage to the lungs by the trachea, and to inflate them with air, to produce artificial respiration, and to *transfuse blood into him from a lamb*."[2]

Thornton's idea of transfusing the dead president was swiftly rejected by Washington's family. They did not quibble with the doctor as to whether resurrection by transfusion could be possible. Instead they declined on the grounds that it was better to leave the memory of George Washington's legacy intact as "one who had departed full of honor and renown; free from the frailties of age, in full enjoyment of every faculty, and prepared for eternity."[3] Death was preferable to any extraordinary attempt to resurrect the president using animal blood.

Thornton was not the first to propose blood transfusion as a miraculous cure, nor was he the first to consider animals as donors. More than 130 years earlier, between 1665 and 1668, all of Europe was abuzz with excitement over the possibility of blood transfusion. French and English scientists were locked in an intense battle to master blood's secrets and to perform the first successful transfusion in humans. Members of the British Royal Society began by injecting any number of fluids into the veins of animals: wine, beer, opium, milk, and mercury. Then they turned their sights on transfusions between dogs—large ones to small ones, old ones to young ones, one breed to another. The French Academy of Science followed suit with its own canine transfusion experiments but to its dismay was unable to replicate English successes.

Then, seemingly out of nowhere, a young physician named Jean-Baptiste Denis surprised the scientific world when he performed the first animal-to-human blood transfusion to great acclaim—and even greater controversy. On a cold day in December 1667, Denis transfused lamb's blood into the veins of a fifteen-year-old boy. The result was stunning: The boy survived. But fate would not be kind to Denis for long. Flushed with success, Denis tried his next, and last, round of transfusions—this time on a mentally ill, thirty-four-year-old man named Antoine Mauroy. The doctor cut open the vein of a calf and rigged a rudimentary system of goose quills tied together with string. He then transfused just over ten ounces of calf's blood into Mauroy's arm. By the next morning signs looked promising that the experiment was going to work—or, at the very least, not be fatal. Several days and several transfusions later, however, Mauroy was dead. And Denis was soon accused of murder.

In a dramatic turn of events, a Paris judge cleared Denis of all accusations on April 17, 1668. Still, the madman's death signaled an end not only to Denis' career as a transfusionist but also to transfusion entirely. In its judgment the French court mandated that no future human transfusion could be performed without prior authorization from the Paris Faculty of Medicine. And this was very unlikely to happen, given that the medical school had made no secret of its hostility toward the procedure. Two years later, in 1670, the French parliament banned transfusions altogether; transfusion experiments were also stopped in England, Italy, and throughout Europe, not to be taken up again for 150 years.

This book views the story of the Denis trial through two different lenses. First, it is a microhistory that traces the little-known and captivating tale of the rise and fall of the transfusionist Denis, and blood transfusion more generally, over a period of about five years during the seventeenth century. But, perhaps more important, it is also a macrohistory that traces the confluence of ideas, discoveries, and cultural, political, and religious forces that made blood transfusion even thinkable in this era before anesthesia, antisepsis, and knowledge of blood groups. This story is, then, as much about the scientific revolution—its greatest minds and most calculating monarchs—as it is about blood transfusion itself.

The term "scientific revolution" has long been a matter of debate among historians, and I should pause briefly here to explain the use of it in my subtitle and at various moments throughout the book. Beginning in the late 1940s with the work of such legendary historians as Alexandre Koyré and Herbert Butterfield, the scientific revolution was understood as the unequivocable birth of modern science—the decisive moment at which science heroically supplanted superstition and never looked back. This is not how I use the term. Instead I join more recent historians who have worked diligently to nuance our understanding of the scientific revolution as, explains Steven Shapin, "a diverse array of cultural practices aimed at understanding, explaining, and controlling the natural world, each with different characteristics and each experiencing different modes of change."[4] If we can be certain about one thing when it comes to the scientific revolution, it is that there were no easy answers, no clear consensus. Natural philosophers—as scientists were called then—tussled with one another to unlock nature's truths, and more often than not they disagreed, sometimes violently.

The early chapters of this book begin across the Channel, in England, where the foundations for the Frenchman's history-making transfusions were laid. Here, men like Christopher Wren, Robert Boyle, and Robert Hooke performed experiments to test William Harvey's recent discovery of blood circulation. This foray into the larger, international context of early science—fascinating and drama-filled in its own right—lays the foundation for what was, in the end, a showdown between France and England in the fight for scientific dominance.

At the core of this battle lay the race to solve enigmas about both earthly and divine worlds that were as complex as they were controversial. In the early decades of the seventeenth century, Harvey turned understandings of the human body upside down when he announced his discovery of blood circulation. René Descartes had also made his radical pronouncement, "Cogito ergo sum [I think therefore I am]," claiming that the mind—and the soul—were independent from the body, which he argued was little more than an ingenious machine. In a Europe still recovering from the ravages of religious wars, natural philosophers tried to make sense of the broad implications of these and other theories that had so unsettled traditional understandings of science and the body.

A perfect storm had been brewing in Europe, between France and England, Catholics and Protestants, and, especially, science and superstition. And transfusion sat at the heart of it all. It is through this larger cultural and political narrative that Denis' experiments must be told. The early story of transfusion is not just about an ambitious man whose efforts met with resounding failure; it is the story of a world undergoing radical transformation as science and society changed at a pace never before imagined.

As novelistic as this world of early science may seem, however, this book is a work of nonfiction. My narrative approach to history frames what is, above all, a study of historical documents, manuscripts, medical manuals, personal letters, and illustrations—some well known and many esoteric, even obscure. Still, to research and to write about history also means coming face-to-face with any number of conundrums, contradictions, and archival gaps. Confronted by those moments, I have relied on what I know intimately about the time period, its actors, and its power structures as a professor of early medicine and culture. Yet, as imperfect as our knowledge about any moment in the past necessarily is, the lines between fact and fanciful speculation must nonetheless be firmly drawn. In this regard I am indebted to cultural historians such as Carlo Ginzburg, Natalie Zemon Davis, Robert Darnton, and David Kertzer, whose scholarship guided my own efforts to make history's stories come alive—responsibly and in ways that breathe life into a chapter of early science that might otherwise be lost to general readers.

History books rarely mention early transfusion, primarily because it does not fit at all neatly into the larger narrative of the "revolutionary" triumphs of science in the seventeenth century. As one scholar wrote, "It is probably fortunate that blood transfusion took a nap for over one and a half centuries. Ignorance of antisepsis, asepsis, and immunology would have resulted in countless disasters."[5] This is likely true. The few historians who have studied early transfusion argue that the procedure was outlawed in France, England, and Italy in the wake of the Denis trial because it was too deadly.[6] And when I first learned about the Denis case several years ago, I was inclined to agree. However, the more I researched contemporary accounts of the experiments and the court case that followed, the less this argument held up.

It does not take an advanced degree in immunology to imagine how transfusing animal blood into human veins could be dangerous, or even lethal. But surely if mortality concerns alone were at the heart of the initial prohibitions against transfusion, many other procedures would also have been banned. Bladder stones, for example, were frequently removed through penile extraction or by cutting so deeply into the perineum that the barber-surgeon could reach his whole hand into the patient's body. The operation was so painful that the Renaissance surgeon Ambroise Paré explained that it took four strong men to hold his patients down during the operation. The procedure was also notoriously fatal—so much so that the English diarist Samuel Pepys put his stone on display and celebrated each anniversary of his own procedure, exceedingly grateful that he was still alive.[7]

Similarly, no formal limits were imposed on what was arguably the most emotionally and ethically fraught of operations: cesarean sections. In the absence of effective anesthesia, cesarean sections were excruciatingly painful and often resulted in the death of a new mother in an act that was meant to save the life of a child whose own survival was anything but certain. In 1668, the same year as the Denis trial, the surgeon Jacques Mauriceau called the procedure "a great excess of inhumanity, of cruelty, and of barbarity."[8] Still, neither the courts nor the medical faculties ever put formal restrictions on these and other horrifically painful and dangerous procedures.

The more I dug into the Denis case, then, the more questions I had—and doubly so after I learned the outcome of the trial. While much of the court record surrounding the Denis trial has been lost to history, all existing seventeenth-century accounts do agree on one thing: Mauroy was poisoned—not by the animal blood that may or may not have flooded his veins—but by arsenic. All accounts also agree that several doctors, whom Denis would later call "Enemies of the Experiment," were directly implicated in the death. But over the centuries, oddly enough, the names of these men have been relegated to the dusty shelves of history. To date no study has attempted to unveil the identities of the doctors who feared transfusion so much that they would resort to murder. Who could they have been? And what could have been their motivations to kill?

The truth, as they say, is sometimes stranger than fiction.

"Enemies of the Experiment" lurked everywhere, it turns out, and their reasons for wanting to put an end to transfusion are as strange as they are fascinating: It was the moral and religious implications of mixing the blood of different species, rather than the medical safety or well-being of the patient per se, that put a stop to the first transfusions.

Some seventeenth-century physicians and power brokers feared that science was toying with forces of nature that it did not understand—and very dangerous ones at that. In early Europe the borders between science and superstition were as fluid as the blood with which natural philosophers were experimenting. Detractors compared doctors who practiced transfusion to alchemists. Just as alchemists worked tirelessly to transmute base metals into gold, transfusionists risked transforming bodies and minds by transfusing animal qualities into human veins. Would humans now bark? Or dogs begin to speak?

In early European minds the potential for species transmutation via transfusion was real—and terrifying. Monstrous hybrid creatures loomed large in the early European imagination. Sea dragons put the fear of God into the hearts of New World explorers; sailors returned from their travels with tales of kingdoms ruled by dog-headed men and islands inhabited by mermaids who were neither fully human nor fully fish. For some the risk that science could create monsters—or worse, corrupt the entire human race with foreign blood—was simply too much to bear. Transfusion needed to be stopped, and it was, for well over a century and a half after the Denis trial.

"Blue blood," "true blood," "blood brothers": In any era blood gets to the heart of who we are, or at least, who we want to believe ourselves to be. Perhaps nowhere is there a better example of the obsession with blood and identity than in the

1940s, when the rhetoric underlying American racial segregation made its way to the blood banks.[9] In November 1941 the American Red Cross—mirroring the social divisions prevalent at the time—announced that it would not accept blood from African American donors for use in its blood banks. Two months later, in January 1942 and in the wake of considerable criticism, the Red Cross agreed to collect and store the blood of "colored" donors. However, the organization also made it clear that the blood would be segregated. In the absence of clear scientific evidence to support their decision, blood segregation appeared to work largely as a way to calm cultural fears of contagion. In one of thousands of letters addressed to U.S. senators and representatives, for example, an anonymous writer expressed concern about what multiracial transfusion would mean for white men returning from World War II. The blood of another race might not have a visible effect on the recipient himself, the writer worried, but it would corrupt the purity of bloodlines for generations to come: "How many white men, having a choice, would rather die there on the battlefield without plasma than run the chance of coming back to be the father, grandfather, or great grandfather of a brown, red, black, or yellow child?"[10]

Such deeply felt debates on blood and race continued to rage for another two decades—and exploded in 1959, when the physician John Scudder and his colleagues presented the case study of a white man who had died, they claimed, from a blood incompatibility reaction following open-heart surgery. By all appearances the patient had received blood from a white person that was perfectly compatible with his own. But the man's death, they argued, could be traced to an earlier transfusion—in which he received blood from an African American donor.

The donor's blood contained an antibody (Kidd negative, JK^a) that researchers believed occurred more commonly in blacks than in whites.[11] The first transfusion caused a reaction to the Kidd-positive antigen in the man's own blood, so that the result was deadly when additional Kidd-positive blood was transfused into the recipient from the second, white donor. "If a white donor had been selected for the first transfusion," the researchers explained, "the chances of our patient receiving Kidd negative would have been three times less than if blood from a Negro donor had been used, as was the case."[12]

The researchers argued for a race-based triage protocol when selecting suitable donors. The best blood, the Scudder group argued, was one's own. If that was not possible, then blood from a twin sibling or a blood-group-compatible family member was best. Still absent this, only "compatible blood from donors of the patient's race" should be used. A year later Scudder and his colleague W. D. Wigle argued for a "new philosophy in blood transfusion . . . 'Unto each his own.' "[13]

Although the United States was in the thick of intractable segregationist social policies, physicians and researchers were vocal in their disagreement with Scudder's claims. In the pages of the *New York Times* seven physicians from Columbia University scolded Scudder and his colleagues: "The so-called 'new philosophy' [of blood transfusion] serves no useful purpose except to reinforce the old 'philosophy' of race prejudice." The newspaper bolstered the Columbia group's message by publishing, on the same page, a report from a South African Red Cross official who explained that interracial blood transfusions had been performed for more than twenty years in that country without adverse effects.[14] And a year later Dr. Eloise Giblett of the Central Blood Bank in Seattle presented a quantitative refutation of Scudder's conclusions at the next American Association of Blood Banks meeting, stating that the data gave no "support to the alleged advantage of intra-racial over inter-racial transfusion."[15] Despite these and other scientific calls for desegregation, however, blood grouping according to race continued well into the early 1970s in several southern states.[16]

The American Red Cross has long since adhered to strict, nondiscriminatory practices in its blood and biomedical services. Relying on the goodwill of over 4 million blood donors each year, the Red Cross is the largest supplier of blood and blood products in the United States. And the need is great. The national organization estimates that every two seconds someone in the United States requires a transfusion. From patients with sickle cell anemia to those suffering devastating injuries, a safe supply of blood can mean the difference between life and death. Through both its disaster relief services and blood services, the Red Cross fulfills daily and in countless ways its mission to provide neutral humanitarian care to those in need.

This book does not explore in depth the triumphs and tragedies of modern blood transfusion. It is, instead, about the blood science in the seventeenth century—its discoveries and its deadly politics. However, as mid-twentieth century concerns surrounding race so clearly suggest, to narrate blood's past is also to reveal the core concerns of a society at a given moment in time. It would be hard, I think, to read this book and not consider the ways in which early medical history is also suggestive of the debates that swirl furiously around biomedical innovation and the contours of human identity today.

As I write this, intense debate continues to brew over whether or not scientists should be allowed to pursue certain lines of research if public response to that research is mixed or even hostile. Perhaps nowhere is this more evident than in the most recent controversies surrounding human embryonic stem-cell (hESC) research. Because hESCs are pluripotent, they have the potential to develop into any of the hundreds of different types cells found in the human body. Pluripotent cells, scientists argue, may one day hold the cure for myriad diseases, from cancer to Parkinson's disease. For others, hESC research—even on embryos slated for destruction in fertility clinics—constitutes an affront to the dignity and sanctity of human life. In late August 2010, the U.S. District Court for the District of Columbia put a temporary hold on federal funding of hESC research. The court cited the 1996 Dickey-Wicker Amendment in its decision, which prohibited the use of federal funds for "research in which a human embryo or embryos are destroyed, discarded, or knowingly subjected to risk of injury or death greater than that allowed for research on fetuses *in utero*." The ruling has been appealed by the Obama administration, and signs look good that the issue will soon find its way to Congress, which will decide whether to discard, rewrite, or let the amendment stand. Conversations, at once measured and frenzied, about hESC as well as other hot-button issues such as cloning and interspecies chimerism echo many of the same tensions between scientific innovation and the cultural and religious "push-back" that took place during the first blood transfusion trials some 350 years ago.

Alan Leshner of the American Association for the Advancement of Science argues that "for many scientists, any such overlay of values on the conduct of science is anathema to our core principles and our historic success. Within the limits of ethical conduct of science with human or animal subjects, many believe that no scientifically answerable question should be out of bounds." However, as Leshner and others also recognize, when research begins to touch on "the essence and origins of human life," it becomes more important than ever that science and society find ways to interact with one another in open and rational dialogue.[17]

The past, I would argue, offers an excellent starting point for discussions about the future. Worries about science and its impact on what it means to be "human" are not the domain solely of our genomic age. Scientific hopes and high-pitched social fears swirling around early blood transfusion suggest in many ways the tensions of our own day. Though blood transfusion is now a commonly accepted practice, its history provides not only insights into past clashes between science and society but also cautionary lessons on how to navigate them. I will return briefly to this idea in the epilogue. For now I simply ask readers to keep two questions in mind as they enter the teeming streets and cluttered laboratories of seventeenth-century Paris and London: Should a society set limits on its science? If so, how and at what price?

BLOOD WORK

Chapter 1

THE DOCTOR AND THE MADMAN

Paris, December 19, 1667

The French capital was cloaked in a dense and sooty haze as Parisians kept fires burning in chimneys and in the streets to stave off the icy gusts of wind blowing across the Seine. The winter of 1667 was one of the coldest on record; warmth had been difficult to come by.[1] The price of wood skyrocketed, fresh food was nearly impossible to find, and survival was a daily challenge for the penniless multitudes that inhabited the lower rungs of French society. In this city of four hundred thousand, death had become part and parcel of the human experience. Corpses—the product of unrelenting cold, hunger, and violence—filled dark corners of the city's labyrinthine streets. Masses of Parisian citizens, near-corpses themselves, were left to huddle together in a desperate attempt to survive.

Jean-Baptiste Denis stepped out of his home on the Quai des Grands-Augustins and nodded with feigned entitlement in the direction of a waiting carriage. Across the river the gothic spires of Notre Dame Cathedral reached into the gray winter skies. The shivering coachman hovered attentively over the short and stocky Denis as the young man settled comfortably into his seat, warmed by heated bricks. It was getting late, and Denis needed time to double-check that all the necessary preparations were in order for his history-making experiment.

The coachman crossed from the Île de la Cité to the Right Bank by way of the Pont-au-Change bridge rather than the Pont-Neuf. Drivers for well-heeled Parisians did what they could to avoid the riffraff that congregated on the city's oldest bridge—the snake-oil dealers and charlatans, switch-and-bait artists, street actors, and bevies of other shady characters. The Pont-Neuf was also the erotic center of the capital. Heavily perfumed prostitutes strolled along the bridge in décolleté dresses by day, and men of all persuasions fulfilled their passions under the bridge by night.[2] It was not a place for an upstanding gentleman to be seen; there were other, more discreet ways to pursue such interests.

Unlike the Pont-Neuf, the Pont-au-Change had long been associated with restraint and nobility. Since the late Middle Ages the original timbered bridge had been the preferred route of French kings, who made their solemn processions from their home at the Louvre to Notre Dame and back. When the wooden bridge burned down in the late 1630s, it was replaced by a beautiful one in stone.[3] And in a city where dirt and grunge coated the streets, the Pont-au-Change still felt crisp and new.

As the carriage bearing Denis rumbled across the bridge, he could not see the river. Five-story homes and shops obscured the view. This was of little matter; the shops were infinitely more interesting than the gray and polluted waters of the Seine. The windows of Mademoiselle de Tournon's exclusive boutique presented an extraordinary display of sparkling brooches, necklaces, and rings; Messieurs Poirion and Vaugon offered discriminating buyers a stunning array of devotional books, meticulously engraved and illustrated by the best artists in the city; and then there was Monsieur Cadeau—literally "Mr. Gift"—whose ornamented sabers and swords made many men in Paris nostalgic for the good old days at the turn of the seventeenth century, when dueling had not yet been outlawed by the king.[4] There was no way that Denis could afford such luxuries, at least for now. Yet he settled into the comfortable velvet interior of the coach that had been sent for him, certain that one day soon all this—and more—could be his.

Unlike the men among whom he now circulated, Denis' birth sometime in the 1630s had gone unnoticed and undocumented. His father had been a man of modest means—just one of any number of faceless artisans who toiled in hot and smoky metal shops to craft the tools of science that others designed. The elder Denis was a *pompier* by training; his specialty was the water pump that was being increasingly put to a wide range of uses: from rudimentary hand-pumped sprayers used in firefighting to the more elaborate systems that powered fountains in royal gardens.[5]

Despite his family's humble origins, Denis had proved himself to be one of those very rare men who could break through France's brutally static class system. Driven and perennially optimistic, Denis was able to insinuate himself into the world of wealth and high society in order to become, against all odds, a doctor. But while he would certainly have been loath to admit it, Denis was still something of a novice. Barely thirty, he had received his medical degree from the University of Montpellier a few months earlier and had returned north to Paris to seek his fame. Once there Denis was characteristically impatient and eager to distinguish himself. His father may have made the tools that put out fires, but the younger Denis yearned to set the medical world ablaze.

Horses' hooves beat rhythmically on the square stones that lined the streets. Denis' carriage pushed forward onto the main road that led toward the city hall and the large square that fronted it. The Place de Grève would become fully synonymous with death during the Revolution of the next century, when the guillotines turned the streets red. But now the open space was congested with the usual daily chaos. It was full of the carriages and pedestrians who fought to traverse the city, from the universities on the Left Bank where Denis now lived to the blue-blooded Marais district to which he aspired. The carriage inched into the bustling mass. The physician could hear the boisterous taunts of coachmen, his own included, as they challenged one another in their common struggle to pass through walls of pedestrians. As many noblemen would have, the bourgeois Denis ignored the sounds of beggars who pounded on the doors of his carriage, hoping for a handout.

Once Denis was clear of the Place de Grève, the undeveloped banks that lined this section of the river came into view. The marshes had proved difficult to build on and provided a surprising vista of farmlike land along the Seine. Haystacks documented the hard work of fall—and the brutal winter that had followed it—as shivering laborers loaded their wooden boats with feed for the horses and livestock that populated the courtyards of homes up and down the river. Yet the quarter that drew its name from these marshes, the Marais, was hardly bucolic. Not far from the fields on the riverbanks, narrow streets teemed with life. Each linked to the next, imposing noble homes—*hôtels particuliers*, they were called—towered over a warren of tightly clustered streets. Here there were no sidewalks; there was no room for them. Despite laws on the books about how far into the streets the walls of these outsized homes could extend, owners and their builders had found clever ways to add an inch here, an inch there, in a city where space was forever at a premium. The streets below were shadowy, roofless tunnels filled with horses, carriages, merchants, vegetable hawkers, flower girls, pickpockets, and courtesans jockeying for room as they expertly dodged dirt and dung. The more cautious pedestrians pressed against the walls in the hope that they would not be knocked down, or, worse, run over by the constant flow of carriages.[6]

Having successfully navigated the rivers of humanity that flooded the streets, Denis' carriage approached the twenty-foot-high, fortresslike doors of Henri-Louis de Montmor's city estate at 79 rue Sainte-Avoye.[7]

Few men had demonstrated a better understanding of the rank and privilege offered by money than Montmor. Born to wealth, he never questioned whether he deserved the luxuries that he lavished on himself and those who surrounded him. His family had been part of the Parisian social fabric for nearly two centuries. Henri-Louis' father, the elder Jean Habert, was the Master of Requests for Henri IV, the illustrious grandfather of the current king, Louis XIV. A lawyer by training, Jean was responsible for preparing documentation for the legal cases that were regularly brought before the king's council.[8] Over time Jean's work expanded substantially to include full oversight of the war treasury, which earned him the name *Montmor le Riche*—and a well-deserved reputation as an embezzler. Ever resourceful, the elder Montmor once famously swindled a certain Gallet, a wealthy financier known for his love of gambling as much as for his love of building elaborate homes for himself. In an effort to steer clear of his vices, Gallet gave all his gaming money—some one hundred thousand francs—to Jean to hold. A desperate Gallet showed up one day on the steps of Montmor's house and begged his friend to give him some of the stash—just a little—promising that he would wager only this small amount and nothing more. "My dear Monsieur Gallet," replied Jean, "you are dreaming. Your losses have troubled your brain. I have nothing of yours." It was said that he confessed the whole episode to his priest but claimed to the very end that his intentions had been pure. The priest, eager to calm the dying man's spirit and persuaded by Jean's show of contrition, confirmed that he had done a very noble deed. He had saved his friend from damnation; it was certainly better that the money went to him rather than to the devil. Jean died shortly afterward, convinced that he was on his way to heaven.[9]

Like his father, Henri-Louis had found ways to take advantage of the great wealth that flowed freely into the family's coffers. As Jean's only son Henri-Louis wanted for nothing. The same doctors who served the king were regularly called to the Montmor residence at the child's slightest sniffle or cough. The honor of serving Lord Montmor, and the generous compensation they could anticipate for their good deeds, had caused a few fistfights among eager physicians.[10] Jean's money also assured the younger Montmor's access to the same royal power brokers who had established his father's career. By the age of twenty-five Henri-Louis had been named councilor at parliament and, at thirty-two, became Master of Requests as his father had been. His work in this position was hardly stellar. As one of his critics at court wrote, "He expresses himself with difficulty and is slow, timid, and does not apply himself."[11] But money and connections were what opened doors, not talent; and Henri-Louis rarely found a door that he could not open.

Denis' carriage pulled up to Montmor's imposing home. The presence of four guards, two on each side, sent a clear message that entry was by invitation only. Yet this did little to keep street folk, especially the beggars in their torn and dirty clothes, from swarming the entrance with every carriage that arrived. As the huge doors swung open, the guards leaped into action. Truncheons in hand, they clubbed those who would make so bold as to infiltrate this sacrosanct space—men, women, children, healthy and infirm alike. The coachman led the horses through the imposing and distinctive porte cochère that served as a transition between the rude streets and grand taste. Once the coach arrived safely inside, a smartly dressed valet in a wool overcoat, tight-fitting trousers, and polished boots opened the carriage door. Bowing, he greeted Denis, who returned the show of respect with a practiced nod.

The valet swiftly guided Denis from the courtyard to the main entrance of the residence. Denis' heeled shoes clicked with force as he scaled the pristine marble risers of the estate's central staircase. He entered the room with confidence, ready for what would be—he was sure—his greatest moment and a certain confirmation of his talent. The vaulted ceilings reflected the warm glow of what looked like a bonfire burning in the enormous stone fireplace. At one end of the room stood the graying Montmor, who held court with a group of elite guests who had been handpicked to witness this latest triumph of medical science. The nobleman's lively blue eyes engaged each of the men with both comfortable familiarity and studied aloofness. Glancing up briefly to find young Denis standing awkwardly near the door, Montmor offered him a warm greeting and confident assurances that all was moving according to plan.[12]

In the center of the room the surgeon Paul Emmerez was emptying his wooden toolbox and carefully placing his surgical instruments on a nearby buffet: knives and scalpels crusted with blood, clamps of various sorts, scissors, thread, muslin drop cloths stained rust brown, and several large bleeding bowls.[13] Just steps away a local butcher was straining to lift a young calf to the large central table with the help of Montmor's stablemen. Moaning, the animal struggled until it was subdued by blows, the men working quickly to restrain the calf on its side.

Then, as if on cue, there were loud shouts, and the room's heavy wooden doors flew open as several watchmen dragged in an unwilling and clearly deranged Antoine Mauroy. The dirty and unshaven man continued to resist, leaving marks from his bare and calloused feet on the cold stone floor as he struggled. The butcher and his team quickly finished their work on the calf and rushed to help shove the screeching Mauroy into a chair. A few quick loops of a rope followed by a tight tug: Mauroy, like a tamed animal, now had no choice but to submit to the gruesome experiment soon to come.

Standing at a distance from the fray, Denis recalled the first time he saw Antoine Mauroy. It had been summer. The madman was stomping ankle deep in the mud on the marshy banks of the Seine. Naked but for the few rags held around his body with straw rope, Mauroy muttered incomprehensibly and reached up frequently to center a tattered little hat on his head. The homeless Mauroy attracted crowds of schoolchildren who followed him along the river. Normally oblivious to the world around him, the man stopped from time to time and stood still for several seconds. Turning his filthy face suddenly toward the children, he would let out a howl and flap his arms wildly. The children ran, shrieking with delight, and Mauroy retreated into his delusions.[14]

Mauroy had been selected for the experiment because he was one of the most famous, or rather infamous, men in the tight-knit community of nobles living in the Marais. Most of the quarter's elite remembered him as the Marquise de Sévigné's perfectly mannered and well-dressed valet, the one who smiled with compassion as they nervously straightened their wigs or tugged at their corsets before entering the *salon* of the exacting marquise.

Now peals of laughter echoed throughout tastefully appointed reception halls as women in ribbon-decked dresses and men in wigs and flouncy cravats exchanged tales of Mauroy's exploits. According to one well-worn story, cavalry guards were making their nightly tour of the Marais. As their horses nipped down into the hay, munching and snorting, they awoke the naked Mauroy, who had settled into the bale for the night. He responded with bansheelike screams; the horses bolted, and the guards swore to anyone within earshot that they had been chased by the devil himself.[15]

Montmor, Denis, and Emmerez felt certain that if they could cure Mauroy, they would soon become as legendary as their patient. And so it was, at six o'clock on that cold December evening in 1667, that the blood transfusion began. Lamps had been lit, and chaotic energy filled the air. A crowd of physicians and surgeons continued to stream into the room—anxious for the show to begin. Pushing back the crowd, Emmerez first drew ten ounces of blood from Mauroy's right arm and then opened the calf's femoral artery. The madman's demands to be released competed with Montmor and Denis' angry shouts to the spectators to back up and quiet down. Emmerez swore as he was bumped and jostled; he was working diligently to unite the two transfusion tubes while trying to avoid a face full of blood. To no small frustration of the transfusion team, only five or six ounces of calf blood made it into the man. Yet Mauroy began to sweat profusely; his arm and both armpits were burning hot. The room began to spin around him.

While the men had no way of knowing this, Mauroy's immune system was launching an attack on the foreign antigens in the calf's blood. Typical symptoms of a hemolytic transfusion reaction include fever, chills, fainting, or dizziness, as well as bloody urine and back or side pain. They begin most often shortly after a transfusion of incompatible blood, either from a human of a different blood group or, as in this case, another species. Antibodies produced by the recipient's immune system attack the donor cells and cause them to burst. The severity of the blood reaction depends on the amount of blood transfused, the rate at which it is transfused, and whether the patient has had previous exposure to the incompatible blood. Yet this knowledge of blood groups and their importance was still three hundred and thirty-four years in the future. It would not be until 1901 that Carl Landsteiner performed what was actually a very simple experiment and noticed clotting in some samples of mixed blood and not others. The Viennese doctor separated his samples into three groups: A, B, and C (what we now recognize as O), according to their clotting tendencies.

Landsteiner initially overlooked the group AB, which occurs in just 3 percent of populations. In 1907 two researchers working independently—Jan Jansky in Czechoslovakia and William Lorenzo Moss in the United States—uncovered this fourth group. They used roman numerals (I, II, III, IV) to designate each blood group. Jansky classified what we now call group AB as IB, and Moss classified it as I. To avoid confusion the American Association of Immunologists adopted, in 1927, at Landsteiner's urging, the now-standard notation of A, B, AB, and O.[16]

What the seventeenth-century doctors could know, however, was that if they did not stop the transfusion immediately, their patient would soon be dead. As Mauroy swooned, Emmerez ripped the small metal tube from his arm and closed up the wound as quickly as he could. The limp and pale man was helped to his feet and carefully accompanied to the servants' quarters to recover. When the room had finally cleared and the help had been called to clean up the calf's carcass and the bloody mess that went with it, all that could be heard were Mauroy's faint whistles and insane rants echoing across the adjacent courtyard. But as the sun rose the next morning, Mauroy appeared to be somewhat less deranged than before—in fact, he seemed to be an altogether changed man.

Denis and Emmerez decided to tempt fate and try a second transfusion. The two men had persuaded Montmor to be more prudent with the guest list, which they limited to a much smaller, better-behaved, and more elite crowd of physicians. Two days later, again at precisely 6:00 p.m., a weakened and more docile Mauroy was led into the room. Barber's blade and bloodletting pan in hand, Emmerez could not find a vein in the right arm. The men speculated that this was no doubt the result of the toll that Mauroy's living conditions had taken on his body. Mauroy had suffered from months of homelessness, hunger, and cold; there was no possibility, they blindly concluded, that his condition could have been the consequence of the earlier experiment. The left arm proved more successful. Two ounces of blood were removed, and more than sixteen ounces of calf's blood took its place—nearly triple what had been transfused into Mauroy during the first experiment.

As soon as the blood began to enter Mauroy's veins, his pulse quickened. He began to sweat in the draft-filled wintry room. He cried that his kidneys hurt, that he was nauseous, that he would choke to death if he was not released from this experiment, this hell. Sensing that they might have gone too far, Denis ordered the tube connecting man to animal be removed. As Emmerez set to work closing the wound, the homeless man promptly vomited the "store of bacon and fat" he had gulped down shortly before and continued to purge "diverse liquors" until he passed out from exhaustion two hours later.[17]

When Mauroy awoke the following morning, he was calm and alert. With uncharacteristic politeness, he requested that a priest come to his bedside so he could confess his sins. After the confession Father Veau closed the door quietly behind him and paused to marvel out loud at what he had just seen. Mauroy was now of sound mind and would actually soon be fit to receive the Sacrament.[18]

As Mauroy continued to rest under the watchful eye of the transfusionist, the madman's wife searched the streets for her missing husband. News of the transfusion had circulated throughout the city and into the countryside. The haggard and penniless woman soon found herself in a home she would never before have dared to enter. Perrine Mauroy slunk toward her husband with great trepidation. She winced as Mauroy leaped from his bed, and she looked surprised, very surprised, when he embraced her passionately. According to Denis' clearly self-interested account of the couple's interactions, Mauroy explained in great detail and "with great presence of mind" to his wife all that had happened to him since she had last seen him: his follies in the street, his naked rants, and—of course—the transfusions the "kind physician" had performed on him.[19]

The wife turned, dumbfounded, toward Denis and stammered a quiet thank-you. At this time of the year and in this "full of the moon," her husband should have been quite insane. Instead of the kindness he was now showing her, she whispered, he would have done nothing but swear and beat her. Madame Mauroy felt both relieved and reluctant to be again at her husband's side. When Denis finally released the former madman from his care, her reluctance turned to fear and dread. The couple returned to their modest, debt-ridden life on the outskirts of Paris. Perrine had spent several comfortable days among the rich and famous; now she found herself once more in poverty, frightened—and wondering when her husband's anger would unleash itself anew.

While Perrine shuddered in fear, Denis proudly set himself to announcing the details of his successes as broadly as possible. In the months that had preceeded this history-making experiment, the transfusionist had perfected his technique using a host of dogs, cows, sheep, and horses. His efforts paid off, and he reveled in the pleasure of his newfound celebrity. Yet it would soon be short-lived. Soon, Mauroy would be dead. And Denis would be staring down accusations of murder.

Chapter 2

CIRCULATION

England, 1628–1665

If Denis had just made a name for himself as France's premier transfusionist, he had France's enemies—the English—to thank for it. For almost four decades English physicians and natural philosophers had tried to make sense of blood's mysteries. The results of their efforts had been stunning. In 1628 the Englishman William Harvey made a discovery that rocked the foundation of medical models that had endured unquestioned for nearly two millennia. His arguments that blood circulated through the body set off a flurry of experiments by men such as Christopher Wren, Richard Lower, Robert Boyle, and Robert Hooke. With each experiment, they took one step closer to attempting transfusions in humans.

In the seventeenth century, research on living humans was, even in the heady years surrounding the early race to perfect blood transfusion, still rare. Instead, medical exploration took place most frequently in the domain of death. Human dissections—which were conducted in university anatomical theaters, public gardens, and private homes—were a regular feature of

European scientific and social life. While natural philosophers had occasionally been known to dissect newly defunct colleagues, their scalpels and saws were most often focused on executed criminals. Long seen as a punishment worse than death itself, the dissection of criminals was officially sanctioned by Pope Sixtus IV in 1482.[1] Fifty years later, in 1537, Pope Clement VII gave formal permission to include anatomical demonstrations, again on criminals, in medical school curricula.

On the mornings of hanging days the church bells rang to let Londoners know that a much-anticipated show was about to begin. Prisoners were led from their squalid cells at Newgate Prison to an "Execution Sermon" in the prison's drafty third-floor chapel. They were seated—men on one side, women on the other—around a coffin as they listened to a priest warn of hell and brimstone, repentance and forgiveness. Spectators, who paid handsomely for the chance to witness the last desperate moments of the condemned, were separated from the sinners by a low wall.[2] Some prisoners pleaded their innocence and begged for their lives. Others were contrite and pledged their souls to God in return for being spared. The most hardened of the "malefactors," as they were called, spit proudly and swore with disdain at their confessors.

This dramatic prelude to the execution accomplished, the prisoners were then led away, shackled together at the ankles, to the open carts that would carry them to the gallows. Once aboard, sitting amid the rough-hewn coffins that housed their futures, they jostled against each other for room. Three long miles separated Newgate Prison from Tyburn, the infamous village where London had staged its executions since the early Middle Ages. A scaffold specially designed for mass executions awaited its next shipment of souls. The "Tyburn Tree" consisted of three posts that rose ten to twelve feet in the air. In an ingenious configuration that allowed multiple hangings at the same time, crossbeams connected each of the posts. The most recent record for simultaneous hangings had been set in 1649, when twenty-three men and one woman swung together in a public spectacle.[3]

And a spectacle it was. Tyburn attracted the city's underbelly. Families and coconspirators of convicted murderers, thieves, and rapists joined riotous crowds of cutpurses and prostitutes. Yet the most notorious players in this grotesque show were the corpse brokers who formed a league of vulture-like wheelers and dealers. They competed for access to fresh cadavers, which they would provide, at a good price, to London's various medical practitioners and medical schools.

William Harvey was one of the many men who covetously sought bodies for research. Harvey was so convinced of—one might even say obsessed by—the utility of dissection that there seemed barely a day that he did not have the body of a human or an animal splayed out in a state of half destruction on one of the large wooden tables in his home. Harvey's belief in "ocular demonstration," as he called it, was relentless. And any corpse was fair game for his exploration. He is even said to have performed postmortems on his own father, sister, and a close friend.[4]

Harvey was among those who argued that it was time to shrug off tradition and blind reliance on the wisdom of ancient writers whose theories had dictated medical practice for millennia. Such theories were not informed by firsthand observations of the inner structures of the human body: The lengthy medical treatises of such men as Hippocrates, Galen, and Aristotle were founded on interpolations made from their work on monkeys, pigs, and other animals. Deviating dramatically from these influential predecessors, Harvey, however, believed that any physician worth his salt had no choice but to roll up his sleeves and get his hands dirty in exploring the mysteries of the human body.

Before long Harvey was putting two thousand years' worth

of Galenic knowledge of the human heart and blood to the test. For the influential second-century physician Galen, blood did not circulate. Instead it made a one-way trip from the stomach to the heart. Venous blood, according to Galen, was the product of food that was "cooked" in the digestive tract and filtered in the liver. The blood coursed from the liver and toward the heart, where the fluid seeped through the heart's chambers through what was believed to be invisible porous membranes. Heat in the body was produced by the heart, whose primary responsibility was to burn blood like kindling in a furnace. Respiration was not, it was believed, responsible for oxygenating blood. Breathing was instead a means to blow off the "smoke" or fumes created by the heart's furnace.

This basic understanding of the heart's fires helps to explain early predilections for bloodletting as a first course of action in the event of illness—and as a preventive measure. A fever was considered to be a sure sign of an overabundance of blood. After all, a well-stoked fire can easily turn into a bonfire if given too much wood or if doused with oil. And by the Middle Ages bloodletting had become the unquestioned first course of action for nearly every human ailment imaginable.[5] As one medieval physician wrote:

> Phlebotomy clears the mind, strengthens the memory, cleanses the stomach, dries up the brain, warms the marrow, sharpens the hearing, stops tears, encourages discrimination [careful decision making], develops the senses, promotes digestion, produces a musical voice, dispels torpor, drives away anxiety, feeds the blood, rids it of poisonous matter, and brings long life. . . . It eliminates rheumatic ailments, gets rid of pestilent diseases, cures pains, fevers and various sicknesses and makes the urine clean and clear.[6]

Put more simply, bloodletting cured all. A belief in the "humors" lay at the heart of the early modern period's attachment to bloodletting. Following in the footsteps of Hippocrates, Galen offered an account of humoralist anatomy and physiology that dominated nearly every aspect of medical theory and practice from antiquity to the eighteenth century. Galenism held that the body was ruled by four different bodily fluids called "humors" and that each carried specific properties. Blood, phlegm, choler (yellow bile), and melancholy (black bile) mixed together in proportions that were specific to each individual. This humoral profile was called the "complexion." Good health was the result of a complexion that was perfectly balanced. Illness descended when one or more humors were out of proportion. Purgings, through emetics and laxatives, gave the body the jump start it needed in order to shed unwholesome fluids and to regain necessary equilibrium.

For centuries blood was regularly coaxed out of bodies by barber-surgeons. The same men entrusted with close shaves and haircuts were also responsible for bloodletting, boil lancing, tooth pulling, and trepanning (skull drilling). They did not receive training in universities or through formalized study of books—which were reserved for physicians who actually had little contact with patients in comparison. Barber-surgeons instead learned their trade through the trials and errors of apprenticeship. The instruments of barber-surgeons were crude; the more menacing tools consisted of saws used in amputations, plierlike devices to remove bullets, and hand-cranked drills for trepanning. And they were used under even cruder hygienic practices. Some barber-surgeons traveled with toolbox in hand, from home to home and from village to village. Others set up shop in more permanent locales, in small street-facing rooms. There was no need to hang a sign; the dripping red rags and barely rinsed pans outside were signal enough of the bloody work within. Modern-day barbershops commemorate the early origins of the profession. Now quaint and certainly less macabre, metal-capped red-and-white-striped poles, displayed prominently outside barbers' doors, evoke the bloody bandages and bowls of earlier days.

The toolboxes of local barber-surgeons included the prerequisite lancets, rags, straps, and blood bowls, but they also con-

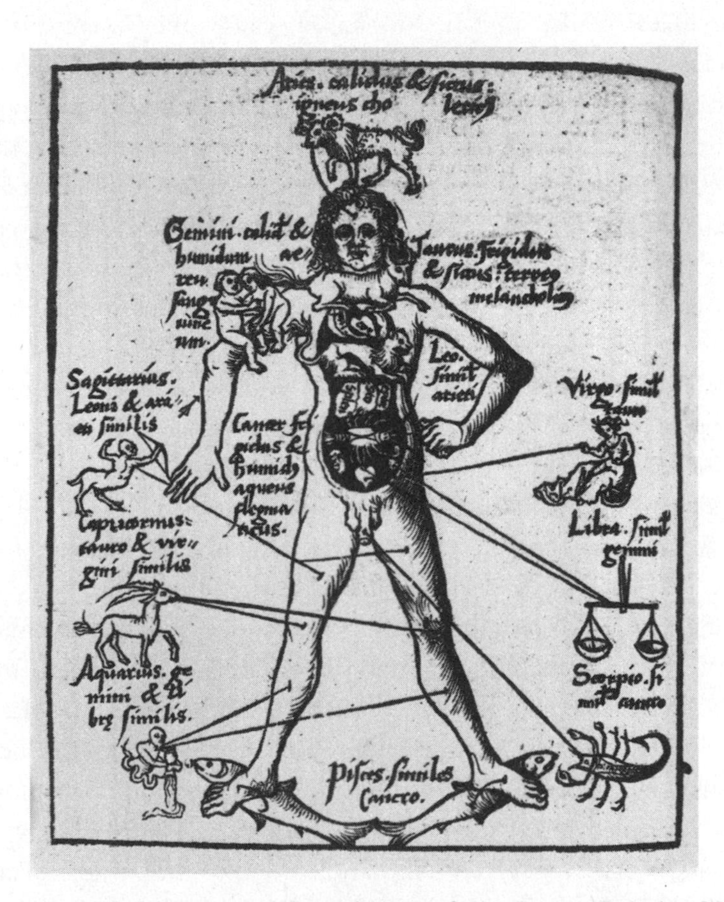

FIGURE 1: Zodiac Men, such as this one from Gregor Reisch's *Margarita philosophica* (1503), were used as easy-to-read guides for the optimal bleeding locations at specific moments of the year.

tained meticulous up-to-date plottings of the constellations. If health was related to the seasons, it was also related to the stars. Astronomy and astrology—the distinction between the two was not evident until the early eighteenth century—played an especially important role in bloodletting. While most bleeding was done from the forearm, bleeding charts showed the places of the body as they were governed by specific star signs. Folded almanacs—called "girdle books" because they were often tucked into belts—depicted the phases of the moon and dates of projected eclipses as well as conventional drawings of astrologically based bleeding points. The heart was connected to Leo; the feet, Pisces; the gut, Libra; and the genitalia, ever-amorous Scorpio. Any bleeding from the body part that matched the current star sign was ill advised. The ailment itself contributed another variable to the complicated calculus required for determining the exact location for bleeding. Writing in the sixteenth century, the celebrated military and court surgeon Ambroise Paré explained, for example, that "a vein of the right arm is to be opened to stay the bleeding of the left nostril" and "a vein is to be opened in the ankle to draw down the menstrual flow in women."[7]

Another method of bloodletting included scarification followed by cupping. Using a multibladed lancet, the barber-surgeon made several shallow incisions close to one another. A small glass cup was heated in a fire and placed over the cuts. While the patient could count on a circular blister from the heat of the glass, the vacuum it created served to draw out the blood from the incisions. Leeches were, of course, another preferred tool for removing unwanted blood from the sick and dying. The trouble with leeches was that they were slimy and hard to control. And, as Paré explained, they could also be fickle. "If the leeches be handled with the bare hand," Paré wrote, "they are angered, and become so stomachful as that they will not bite." He recommended that they be held instead in a white and clean linen cloth. To tempt the leeches to latch on, the patient's skin should be first lightly scarified or "besmeared with the blood of some other creature, for thus they will take hold of the flesh, together with the skin, more greedily and fully." Salt and ash were used to coax the animals to let go, although they were usually left to feast on their hosts until they could imbibe no longer and detached of their own accord. Unlike surgical bloodletting, where bowls were used to catch the fluid and could be used for measurement, it was difficult to know with certainty how much a leech ingested. In these cases Paré made the following recommendations: "If any desire to know how much blood they have drawn, let him sprinkle them with salt made into powder, as soon as they are come off, for thus they will vomit up what blood soever they have sucked."[8]

Bloodletting fell slowly out of favor in the nineteenth century. The work of such men as Louis Pasteur and Joseph Lister ushered in the new understanding that disease was caused by germs and not humors. Their theories were accompanied by a renewed emphasis on evidence-based research and practice. In 1835 the French doctor Pierre Charles Alexandre Louis—often credited with sparking the development of epidemiology as a field—interviewed more than two thousand patients at the Paris hospital La Pitié, recording their autopsies in cases of eventual death. He asked patients when they first became ill, how their disease progressed, and what treatment had been performed. He used this data to assess bloodletting and, while not condemning it entirely, concluded that the usefulness of the procedure was "much less than has been commonly believed."[9] And by the beginning of the twentieth century, bloodletting had moved from a two-thousand-year-old universal intervention to an odd artifact of medical history.

While the early medical world still clung to Galenic humoralism and bloodletting, a number of discoveries set off a chain of questioning about how blood was made and how it moved through the body. In the sixteenth century a little-known anatomist named Amatus Lusitanus speculated that valves in the veins—which he called *ostiola* (little doors)—may in some way direct blood flow, preventing its reflux. This hypothesis, now known to be correct, was promptly dismissed by Andreas Vesalius, one of history's most celebrated dissectionists.[10] Vesalius argued instead that the main purpose of valves was to strengthen the walls of the veins. With Vesalius serving as something of a last word, interest in valves lay dormant until 1603, when the Italian physician Hieronymus Fabricius rejected the Vesalian idea that valves worked merely as architectural reinforcements and returned to the idea of "little doors." He likened them to floodgates, which help control flow volume. Without valves blood would stream unchecked into the lower portions of the body, leaving the upper body parts malnourished. In light of these conflicting theories, valves raised more questions than they answered for the seventeenth-century Harvey. There were so many valves, and in so many places in the body, that he wondered why it was that they "were so placed that they gave free passage to the blood towards the heart, but opposed the passage of the venal blood the contrary way." Surely, he speculated, "nature had not plac'd so many valves without design."[11]

Harvey knew that the only way he could explore his hypotheses on the valves was to perform surgical experiments on live creatures. For as fascinating as human anatomy was and remained for natural philosophers, cadavers presented an intractable problem: They were dead. Dissections had been useful, but Harvey knew that he needed to *see* the blood in motion, to trace its flow through the body, through the valves, and to feel the pulsations of a beating heart. This could only be accomplished through vivisection, by putting living animals under his knife.

Scores of dogs, cats, and pigs roamed the streets and were easily lured with a handful of food; and Harvey kept himself busy with the bounty. Surgery after surgery, he tried to move quickly enough to catch the heart and blood in action. But the animals' writhing was difficult to control, and their heartbeats were too fast. Harvey then turned his attention to cold-blooded creatures. Their slow-beating hearts made eels, snakes, and squid more cooperative subjects. With each heartbeat he plotted the heart's contractions and relaxations; its diastolic and systolic motion. He watched as the heart reddened ever so slightly as it tensed, filled with blood, and then blanched as it forced out its contents. The vivisectionist stared in fascination as his cold-blooded subjects slipped toward death.

Integrating dissection with his observations during live experiments, Harvey was able to quantify for the first time in history

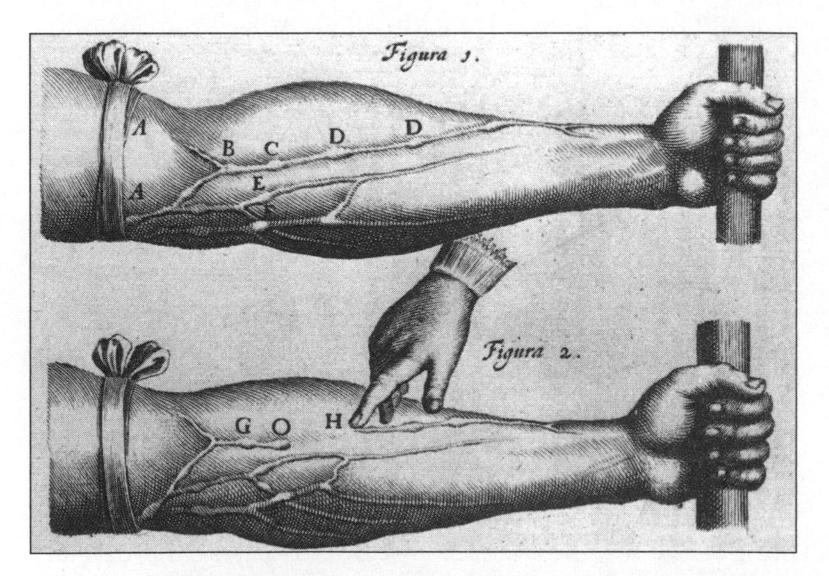

FIGURE 2: William Harvey's illustration demonstrating the action of the valves in *De Motu cordis* (1628).

the amount of blood that coursed through a body. He emptied out all the blood in the chamber of a dissected human heart and determined that he had collected about two ounces of blood. From this he estimated the total fluid that pushed in and out of the heart with every contraction. He multiplied this amount by the number of heartbeats per every half-hour, which allowed him to calculate that nearly 540 pounds of blood would have to be produced and burned off in a Galenic physiological model.[12] This was entirely impossible. Another explanation had to be found. For Harvey it was soon clear that blood did not make a one-way trip to the heart to be incinerated. Instead blood was pumped through the body by the heart in a circular fashion, with the valves helping to direct the flow.

In the decades that followed, many of England's most promising minds spared no effort as they worked to confirm Harvey's claims. For young men like Christopher Wren, Harvey's ideas on blood circulation fitted nicely into what was a much larger fascination with novelty and invention. Even as a teenager Wren had made a name for himself for being as clever with his hands as he was with his mind. Working with fellow inventor William Petty, he devised a machine that easily sowed seeds by drilling a hole in the soil. Wren also ingeniously developed a prototype of a double-writing machine in which two pens were mounted on a frame and could be moved simultaneously to produce duplicate copies of a single document. In this age when letters were the primary form of written communication, such a device promised to be invaluably useful. But to Wren's dismay Petty took full credit when it was presented to Oliver Cromwell—now at the helm of the so-called Commonwealth, following the trial and execution of Charles I—at the end of 1650 or the beginning of the following year.[13]

In 1656 the twenty-four-year-old Wren made a seat for himself at the dissection table and decided to test Harvey's description of the ultimate machine—the circulatory system. In the sixteenth chapter of *De motu cordis,* Harvey had listed other facts that supported experimental evidence for his arguments on the circulation of blood. The most compelling, Harvey argued, was the work of poisons and medications. How was it that a wound from a mad dog could be healed, he asked, but "a fever and other horrible symptoms" could still persist? He concluded that the contagion is carried through the bloodstream to the heart and from there the poison circulates through the rest of the body.[14] With this in mind Wren injected the veins of a dog with wine and ale. The dog soon became noticeably drunk. In an effort to reverse the effects of the alcohol, the precocious Oxford student injected two ounces of an emetic (*crocus metallorum*). The dog "immediately fell a vomitting, & so vomited till he died."[15]

Fueled by excitement about his research, Wren happily bragged to the influential John Wilkins and his friend, the chemist Robert Boyle, that he could easily and quickly convey any liquid poison into the entire bloodstream of an animal. Boyle soon called Wren's bluff by presenting him with a large dog. Unfazed and self-assured, Wren wrangled the dog and strapped it tightly to a table with the help of two colleagues. He exposed a large vein in the dog's hind leg and tied it off. He then made an incision in the blood vessel. Despite the animal's "tortur'd violent strugglings," Wren slid a small grooved plate, which he had made himself, under the vein to hold it in place. He then inserted a thin pipe into the vein. Boyle described at length Wren's next steps and their marvelous effects: "And accordingly our dexterous Experimenter . . . conveyed a small Dose of the [opium] Solution or Tincture into the opened vessel. . . . It was quickly, by the circular of that, carried to the Brain, and the other Parts of the Body: So

FIGURE 3: Early infusion experiments in animals tested Harvey's theories on circulation and laid the groundwork for the first canine-to-canine blood transfusions. Johann Sigismund Elsholtz, *Clysmatica nova* (1667).

that we had scarce untied the Dog . . . before the opium began to disclose its Narcotick quality."[16] Once on its feet, the dog began to "falter and reel." The animal looked so drugged that, wrote Boyle, spectators offered wagers that the dog would soon expire. But to the surprise of everyone, perhaps even Wren himself, their subject not only survived but also grew fat. The dog, made famous by Wren, was stolen not long after.[17]

In 1657, the year of William Harvey's death, Wren moved to London to take up the prestigious post of Gresham Professor of Astronomy. His appointment did little to distract him from his medical experiments. Pairing up with Dr. Timothy Clarke, a fellow Oxford anatomist recently transplanted to the capital, Wren continued his work on infusions. Together they tried injecting "many different kinds of waters, beers, milk, whey, broths, wines, alcohol, and even blood itself."[18] Wren and Clarke then shifted their trials from dogs to men. In the fall of 1657 the infusionists met at the home of the French ambassador to the Commonwealth, the Duke of Bordeaux. The duke, Wren explained, offered up an "inferior Domestick of his that deserv'd to have been hang'd."[19] We have few details on how Clarke and Wren persuaded—or forced—the servant into participating in their experiment. But we do know that Wren in particular was visibly rattled by the outcome. The minute a small amount of the emetic *crocus metallorum* hit the servant's veins, the man fainted. And both Clarke and Wren resolved never again to try "so hazardous an experiment" on humans.[20]

The experience must have left a lasting mark on Wren; he did not attempt medical experiments again in any regular way. But circulation would never be far from his mind—or far from those of other experimentalists who were much less reluctant to impose their dangerous procedures on animals, and soon the bodies of fellow humans.

Chapter 3

THE AGE OF VIVISECTION

The seventeenth century is sometimes referred to as the "Age of Vivisection," and for good reason. The use of live animal subjects was encouraged by the rise of Cartesian philosophy, which held that human and animal bodies were fundamentally similar—because each functioned essentially like a machine. "It is nature," wrote Descartes, "which acts in them according to the disposition of their organs, as one sees that a clock, which is made up of only wheels and springs, can count the hours and measure time more exactly than we can."[1]

As little more than a collection of tubes, pumps, pulleys, and levers, therefore, animals were incapable of language, emotion, and reason. Descartes argued adamantly against critics who claimed that animals found ways to communicate with humans. The philosopher emphasized instead that animals like parrots and magpies may "utter words just like ourselves," but they "cannot speak as we do, that is, so as to give evidence that they think of what they say."[2] While animal lovers might think that their pets expressed pleasure and pain, these were only dispassionate responses to external stimuli. And because animals do not benefit from this capacity of understanding, Descartes concluded, they cannot feel pain. Descartes' arguments on the beast-machine were taken by some as full license for cruelty. After being accused of kicking a pregnant dog, for example, the late-seventeenth-century French philosopher Nicolas de Malebranche responded cavalierly: "So what? Don't you know that it has no feeling at all?"[3]

Vivisection offered new opportunities that many natural philosophers welcomed wholeheartedly, especially members of England's Royal Society. Established in 1660 by order of King Charles II, the Royal Society endeavored to follow the notions set forth in Sir Francis Bacon's *The New Atlantis* decades earlier. It was a state-sponsored "Solomon's House," after the biblical intellectual leader and builder of temples. The society dedicated itself to "promoting physico-mathematicall experimentall learning." Meetings took place once a week at Gresham College in Bishopsgate and featured the most renowned natural philosophers of the day, including founding members Wren and Boyle.

In the first four years of the Royal Society at least ninety experiments were performed on live animals.[4] This figure does not take into account, of course, the many other experiments that were also performed ad hoc on tables in members' homes. One of England's most aggressive vivisectionists was Robert Hooke, known for his observational prowess with the microscope as well as for being the first to use the term "cell" in biology. In their younger years, Hooke and Robert Boyle—two founding members of the society along with Wren—invented a "pneumatick engine" that could be used to create a vacuum chamber. Curious about the conditions within the vacuum, they subjected larks, sparrows, mice, cheese mites, ducks, and cats to the airless horrors of the vacuum.[5] Each of the animals was taken just to, or well past, the point of death before air was once again allowed to fill the chamber. Justifying his research in a Christian context, Boyle brushed off criticisms about the cruelties of animal experimentation in his studies of air pressure: "It is no great presumption to conceive that the rest of the creatures were made for man, since he alone of the visible world is able to enjoy, use, and relish many of the other creatures, and to discern the omniscience, almightiness, and goodness of the author in them."[6]

These early air-pump experiments led to the more gruesome vivisections that Hooke conducted on the lungs of living animals. In 1664 the Royal Society dispassionately recorded that the experimentalist wielded his knife on a dog "and by means of a pair of bellows, and a certain pipe thrust into the wind-pipe of the creature, the heart continued beating for a very long while after all the thorax and the belly had been opened." But mechanistic theories of physiology notwithstanding, even Hooke found this experiment too troubling to repeat. In a letter to Boyle, his partner in research, Hooke described in painful detail the bellows procedure and the "torture of the creature" on which he was experimenting:

> The other Experiment (which I shall hardly, I confess, make again, because it was cruel) was with a dog, which, by means of a pair of bellows, wherewith I filled his lungs, and suffered them to empty again, I was able to preserve alive as long as I could desire, after I had wholly opened the thorax, and cut off all of the ribs, and opened the belly. . . . My design was to make some enquiries into the nature of respiration. But I shall hardly be induced to make any further trials of this kind, because of the torture of the creature; but certainly the inquiry would be very noble, if we could find a way so to stupefy the creature, as that it might not be sensible.[7]

Cartesian arguments that animals were little more than soulless machines had been convenient for late-seventeenth-century natural philosophers and vivisectionists who were clamoring to understand the mystery of the body. But as the reaction of the normally cold-blooded Hooke suggests, Descartes' arguments could go only so far in justifying the obvious suffering that animals experienced while being held hostage to man's scalpels. First articulated in the 1630s, not long after Harvey's game-changing discovery of circulation, Cartesianism represented a radical departure from long-standing Aristotelian ideas regarding the body and the soul. According to Aristotle, the lowest entity on the "Great Chain of Being" was plant life, which possessed a corporeal "vegetative soul" endowed with only the basic faculties necessary for life: nutrition, growth, and reproduction. Higher up, animals enjoyed both this vegetative soul and a "sensitive soul," which allowed for sensation, movement, and—at least to some degree—emotion. Humans alone possessed an "intellective soul," along with the vegetative and sensitive faculties. The intellective soul provided the faculties of knowledge, memory, will, and reason. In a word, humans had minds (*mens*).[8]

For most, mind and soul were embodied. According to the Old Testament the soul was part of the blood itself. For Galen it resided in the liver, which was believed to be the seat of blood production. In later Christian doctrine the soul moved from the blood to the ventricles of the brain, where it was better protected from corrupt, earthly forces. Floating in the dark and empty spaces of the brain, the soul thus inhabited the body but was not *of* the body.

When Descartes evicted the soul from the body, he was the target of hostile critiques from nearly all corners of Europe. Yet the philosopher himself struggled to counteract the objection that thought and emotion can be manifested physically: The act of thinking makes brows furrow, anger tightens the chest, sadness brings tears, desire warms the body. In his later writings Descartes asserted that the pineal gland, nestled (as he understood it) in the center of the brain, served as a way for a disembodied soul to communicate with the body. The movement of the mind on the pineal gland excited the "animal spirits," which then communicated messages to other parts of the body. "The soul has its principal seat," wrote Descartes, "in the middle of the brain. From there it radiates through the rest of the body by means of animal spirits, the nerves, and even the blood."[9] Descartes concluded that, while having no material existence itself, the soul radiates through the body via the pineal gland in the brain.

As vivisectionists continued to perform blood experiments,

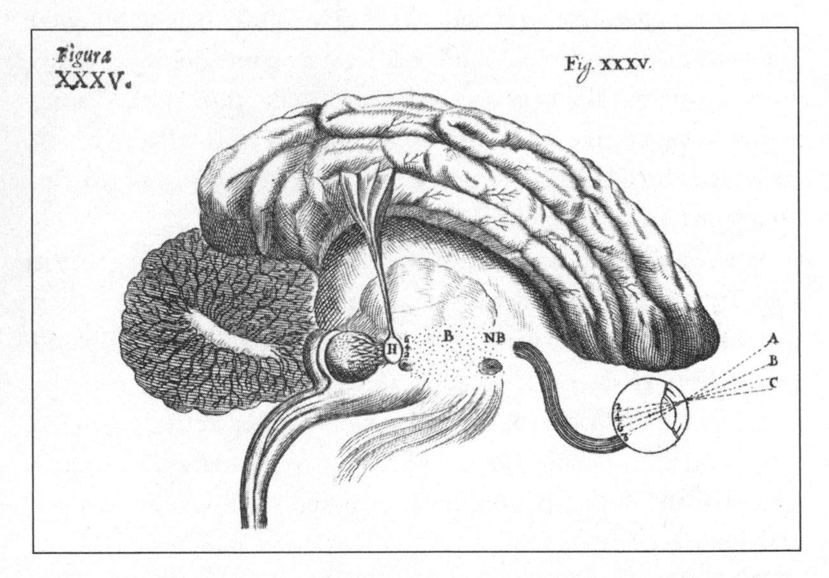

FIGURE 4: For Descartes, the pineal gland (H) helped mediate communication between the corporeal body and the noncorporeal intellect and soul. *De homine* (1662).

they were forced to lay their cards openly on the table. Was the soul corporeal? Did it reside in the blood? What if both animals and humans had souls? And most troubling of all, what if animal and human blood were to be mixed? Over the short span of four years, between 1665 and 1669, precisely these questions would determine the French transfusionist Denis' fate—as well as that of blood transfusion more generally—in both England and France.

Blood experiments were not only a matter of philosophy; they were also a performance that showcased a surgeon's skill in unpacking nature's mysteries. With nimble hands and unbreakable concentration, the surgeon Richard Lower became legendary for his perfectly choreographed surgical displays. It was through his work that blood—and later, blood transfusion—would rise to take medicine's center stage both in England and, not long after, in France. Born three years after Harvey published *De motu cordis,* the sunken-eyed and impassive Dick Lower had earned a name for himself through the dexterity and care he took in his anatomical and physiological experiments. Unlike many surgeons who were known to carve up cadavers—both human and animal—with about the same art as a local butcher, Lower worked slowly and patiently as he chiseled, like a sculptor, through the mysterious flesh of his subjects. He was a dedicated, even obsessive, anatomist who seemed incapable of separating himself from his work. The antiquarian Anthony Wood claimed that Lower often skipped Mass in favor of dissection; indeed, Wood had seen him hard at work on a calf's head on a Sunday morning, in his dissection rooms adjacent to Christ Church college.[10] Even Lower's pets were unable to escape his knife. Another contemporary, John Ward, noted in his diary that Lower owned "a dog which they call Spleen because his spleen was taken out." The dog was, of course, promptly dissected when it finally died about a year later.[11]

Lower's gifts as a vivisectionist surgeon were lauded by Thomas Willis, his professor at Oxford, who acknowledged his student and assistant Lower in *Cerebri anatome* (1664) as "a doctor of outstanding learning and an anatomist of supreme skill. The sharpness of his scalpel and of his intellect . . . enabled me to investigate better both the structure and functions of bodies, whose secrets were previously concealed." Not a day passed without Willis and Lower undertaking "some anatomical administration" on the brains and bodies of a whole bestiary of creatures: "horses, sheep, calves, goats, hogs, dogs, cats, foxes, hares, geese, turkeys, fishes, and even a monkey."[12] Credited with the discovery of the circle of arteries that supply blood to the brain (the circle of Willis), Willis enlisted Lower to help him perform a wide range of infusion experiments in order to track the path of the blood from the brain through the rest of the body and back. A fellow student at Oxford with Wren, Lower injected milk into the veins of dogs and ink into their brains as well as other "kinds of liquors, tinctured with saffron, or other colours . . . to try how the blood moves, and how the tincture may be separated in the brain."[13] A nimble and creative thinker, Lower explored the implications of his teacher's work. Building on Willis's discovery of the arterial circle, Lower established that cerebral circulation could be maintained even if one or more parts of the circle became blocked or narrowed.

Cartesian dualism of mind and body did not sit well with Willis. His many dissections had shown that both humans and animals had pineal glands. This in itself had given Willis plenty of reason to doubt the French philosopher's already tenuous claims of mind-body dualism. Following along the lines of one of Descartes' main detractors, Pierre Gassendi, Willis instead made the case that man was a "two-soul'd animal." Like animals, man had an embodied "sensitive soul," which was responsible for lesser faculties such as growth and sensation and was present in all parts of the body, including the blood. The rational soul, on the other hand—the soul that thinks, feels emotion, and reasons—was also embodied, and located exclusively in the brain. In contrast to Descartes as well, Willis believed that animals did have souls. They showed evidence of memory, decision-making ability, and emotion—which meant that they must have some soul, albeit a primitive one. Yet it was humans alone who benefited from the much-more-complex rational soul.

For Lower debates and questions surrounding the exact nature of the soul were interesting but apparently not of immediate concern. Moving from the brain to the blood, Lower continued Wren's earlier infusion experiments and focused on the possibilities of intravenous feeding. Lower wondered whether he might keep a dog alive "without meat, by syringing into a vein a due quality of good broth, made pretty sharp with nitre, as usually the chyle tastes." Perhaps he could even implant permanently a tube into the animal, so that he would not need to make a fresh incision every time. With this in mind he injected a dog with warm milk; the dog died one hour later. When he later dissected the animal, he found that its blood had mixed with the milk "as if both had curdled together." Like oil and water, there were some things that just did not mix well with blood, he concluded.[14] Not one to shy from a challenge, he wondered aloud in a letter to Boyle if the problem of intravenous feeding might be solved by mixing blood with blood. "As soon as I can get two dogs of equal bigness," he wrote, he would bleed an artery of one dog into the vein of the other "for an hour's time, till they have whole changed their blood."[15]

As the work of Willis, Wren, and Lower demonstrates, Harvey's discovery of blood circulation in the late 1620s set off a chain of questions and experiments that, with the benefit of historical hindsight, clearly set the stage for transfusion. Still, important and unresolved issues regarding the exact nature of blood—especially questions of the exact location of the soul and the extent of animal and human differences—would continue to linger in the background. It would take another year or more for these questions to return front and center on the biomedical stage, but when they did the results would be deadly. In fact this showdown on matters of the soul might have taken place earlier had Lower been able to push forward with his next experiments as he intended. But, just as he was getting started, a devastating plague and the Great Fire of London got in his way.

Chapter 4

PLAGUE AND FIRE

The heavens glowed with a streak of heat as a comet blazed across the night sky. It arrived suddenly one winter evening in 1664 and shimmered across a dark background of stars for nearly two months. The fiery globe loomed over Europe as if taunting astronomers to chase it. In a celestial game of hide-and-seek, the comet first revealed itself to Spanish stargazers on November 17. It peeked out of the gray Dutch clouds on December 3, and on December 14 the comet made itself known to the astronomer Johannes Hevelius at his home in Poland. Then it made a mandatory stop at Cambridge so that Isaac Newton could have a glimpse of its striking long tail. In an order that felt as if the stars were playing favorites, the comet was at last spotted in French skies several days later and soon reached its maximum brightness on December 29.[1]

Once the comet appeared its presence was something of a beacon. It was visible to the naked eye, and its ostentatious hovering struck dread and foreboding in the hearts of those who looked up fearfully toward the glowing light. "Bearded stars," as Aristotle called them, were harbingers of doom.[2] They announced the coming of any number of calamities: drought, famine, earthquake, flood, economic disaster, war, plague, and even the Second Coming. In the comet's wake cathedral bells chimed, priests found new converts, and parents kept their children under a more watchful eye. "Death comes with those celestial torches," wrote one Stoic poet, "which threaten earth with the blaze of pyres unceasing, since heaven and nature's self are stricken and seem doomed to share men's tomb."[3]

The longer the comet lingered overhead, the longer the impending suffering would last. Since the ancient astronomer Ptolemy, star watchers had posited elaborate correlations between the length of future disasters and the time that the comet remained in the sky. And some cometary prognostications were dire. William Lilly, author of *England's Propheticall Merlin,* warned his readers that a year of disaster would ensue for every day a comet was observed.[4] Little wonder that the whole continent breathed a sigh of relief when the November 1664 comet sputtered slowly out of sight two months after it began haunting Europe.

If the superstitious read on the 1664 comet was dim, the scientific view of these astronomical anomalies was strangely bright. In January 1665 the College of Clermont in France held an unusual conference to discuss the current state of comet research and to put forth the next steps for study of the composition, trajectory, and origins of comets in anticipation of when another one would make its dramatic appearance. They did not have to wait long. In March a new comet swooped into view. Dread turned to full panic among the populace—and for good reason. "This comet," declared astronomer John Gadbury, "portends pestiferous and horrible windes and tempests."[5] How right he was. In just a few short months the comet made good on its horrific promises. In

April 1665 a pervasive viral hemorrhagic fever—the plague—swept through London.

Celestial shows aside, London was an ideal setting for such a disaster. The city's tight medieval streets had long been pushed to their limits. Inhabitants elbowed and jostled one another in a futile attempt to claim personal space. Too wide for the narrow passageways, carriages scraped facades of wooden buildings, leaving lines of splintered scars in their wake. Peace and quiet had been nearly impossible to find within the city walls and, even where they could be found, required nothing short of a king's ransom to enjoy. As the diarist John Evelyn complained, "As mad and loud a town is nowhere to be found in the whole world."[6]

If a relentless din battered the ears, stench assaulted the nose. Dogs and pigs roamed the streets freely. Raw sewage flowed everywhere but the "kennels," or channels, dug in the ground to direct rainwater and effluvia. (For obvious reasons sturdy foot scrapers were installed beside every entrance door.) The city was, Evelyn wrote, wrapped in "clowds of Smoake and Sulpher, so full of Stink and Darknesse." Slaughterhouses coexisted alongside family residences; candlemakers filled the air with the smell of putrid tallow. Sparing no criticism for his home city, Evelyn explained that even churches offered little respite; "coughing and snuffing . . . and barking and spitting" were "incessant."[7]

Illness was part of daily life in London, and the capital was ripe for the devastating plague that struck in April 1665. It began slowly enough, with just a few deaths in the city's outlying communities. But in mid-June it hit London with full force. And by the fall of 1665, nearly one hundred thousand people, 20 percent of the city's inhabitants, were dead. Another two hundred thousand fled to the countryside to escape the pestilence. Of those more than 50 percent would soon be struck as well.[8] With the city's mazelike streets now emptied of their usual crush of noise

FIGURE 5: Images of the London plague (1665). From left to right, top to bottom: illness at home, looting, Londoners in boats on the Thames fleeing from the city, Londoners refused entry to villages in countryside, death carts and burials of plague victims in the capital city, ending in the return from the countryside to London.

and dust, London had become a ghost town. There was "a dismal solitude in London-streets," wrote the Reverend Thomas Vincent, "a deep silence almost in every place . . . no rattling coaches, no prancing horses, no calling in customers, nor offering Wares; no London cries sounding in the ears; if any voice be heard, it is the groans of dying persons, breathing forth their last."[9]

Bloated bodies were thrown into open plague pits. On those corpses that had been retrieved shortly after death, the "buboes"—the plague's signature swellings at the neck, under-

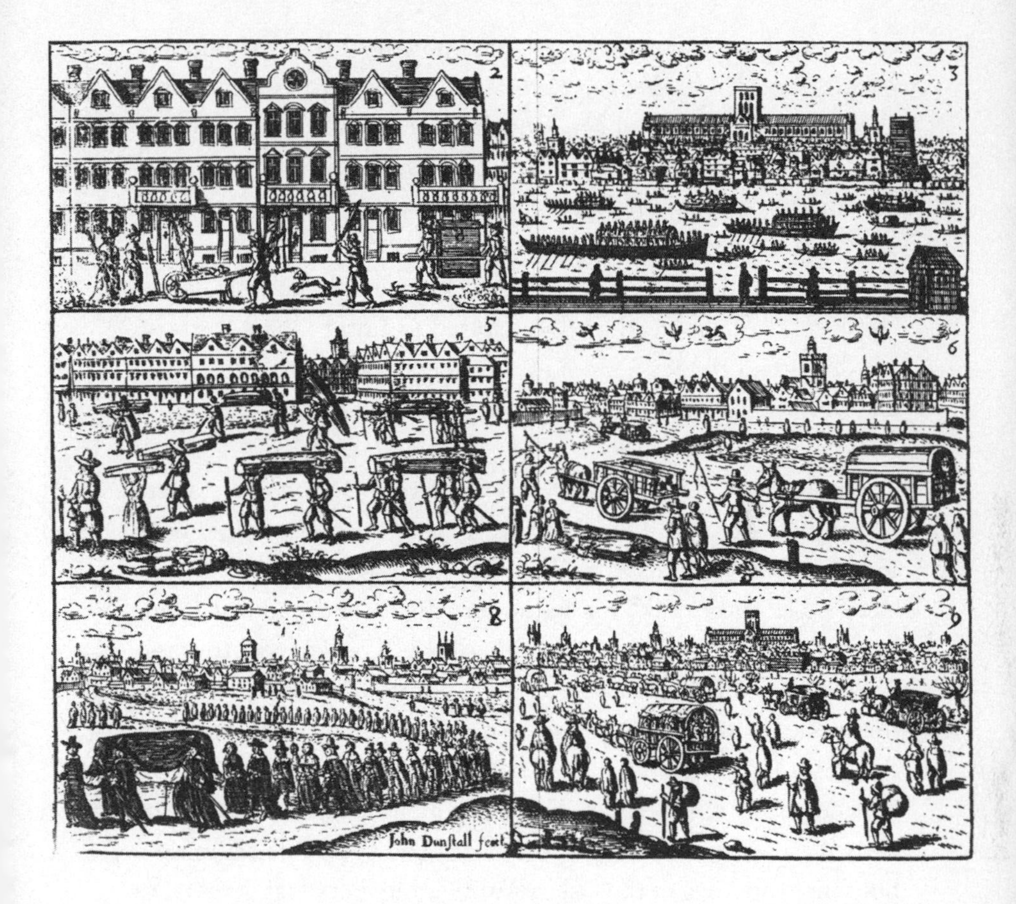

arms, and groin—were still visible. In the heat of the approaching summer others had decomposed into a reeking stew of death that looked anything but human. Unable to keep up with the passings they intended to mark, church bells rang unceasingly through the city's parishes, sending shudders through the few souls who still remained there.[10]

In the month of September alone, at the height of the plague, bills of mortality confirmed that nearly 7,000 of the 8,252 bodies taken away in "dead carts" were plague victims.[11] How the disease spread was anyone's guess. We know now that the bubonic plague is caused by *Yersinia pestis*, a bacterium that is transferred to humans through fleabites. But the seventeenth century would have had no idea of the scourge's origins; germ theory was still more than two hundred years away. Instead miasma—corrupt air—was thought to be the primary suspect.[12] Rotten food, musty rooms, flooded fields, the exhalations of the sick, and the decaying corpses of the dead—all conspired to bring illness to those who breathed the fetid air. And in an era where "horrid stinks" and filth claimed every corner, city dwellers were rarely treated to "wholesome" air.[13]

Army and city officials ordered that fires be lit on every street in a frantic attempt to clear the air. Plague doctors covered neck to toe in black garments walked the city. Their heads and faces were veiled in beaklike masks under which they could be heard hacking and coughing. At the ends of those beaks incense smoldered, in an effort to protect their wearers from the noxious miasma of pestilence. Members of this sinister aviary set upon the suffering city, and they crossed thresholds of homes that no one else dared to enter. Through the glass-eyed openings of their masks, they confirmed that death had staked its claim on a household and began the task of fumigating. Personal items were set afire, the body collectors were called, and the exterior of the house was marked by a double cross: the plague cross.

By early November 1665 the curse began to lift, and by January 1666 streams of exiled Londoners returned home on foot, tired and traumatized. London was still grieving its losses, but life in the bustling city gradually returned to some semblance of normal. And as the streets slowly filled again with people, so too did Lower resume his blood experiments. One wintry February day in 1665 the ever-serious Lower strapped two dogs onto a table. "I tried," he later wrote, "to transfer blood from the jugular vein of one ani-

FIGURE 6: Typical apparel of plague doctors and other officials who entered plague-stricken zones in the seventeenth and eighteenth centuries.

mal to the jugular vein of a second by means of tubes between the two." But no blood flowed; it clotted "at once" in the tube. Lower quickly emptied the tubes and tried to reinsert them another way, again with no success. "I finally determined to transfer blood from the artery of one animal into the vein of a second."[14] Lower later positioned two dogs—the neck of one to the neck of the other—

on a table and began by delicately exposing the carotid artery of the donor. He tied a loose knot around the artery in two places, one above where he intended to draw blood and one below. Then slipping two threads under the artery, he lifted it up ever so carefully so as not to stretch or strain the blood vessel. Scalpel in one hand, a large quill in the other, Lower crouched over the writhing animal. Steadying himself, the surgeon quickly nicked a small opening in the artery and, in one deft move, slid a quill into the pulsing vessel—and then hurriedly attended to the recipient. There, he performed the same choreographed sprint, this time on the jugular vein. Into the recipient he inserted not one but two different quills. The first would be used to connect the recipient's vein to the donor's artery. The second would allow the dog's own blood to be emptied out into a shallow dish.

Moving swiftly and quietly, Lower united the quills and released the bottom slipknot in each dog. His eyes followed the red path of arterial blood as it streamed through and dripped around the quill. The experiment was working. As he watched with his trademark dispassion, Lower allowed the rivers of blood to flow just "till the [donor] dog began to cry, and faint, and fall into convulsions and at last dye by his side."[15] Expressionless as ever, the surgeon turned his full attention to the survivor. When the "tragedy was over" he deftly removed the awkward apparatus from the dog's wound and stitched up his patient. Once freed, the dog immediately leaped up from the table, shook himself, and ran away "as if nothing ailed him." Blood flowed from one dog into another without, it seemed, ill effects on the recipient.

In June 1666 Robert Boyle masked his eagerness with diplomatic reserve as he wrote to Lower, on behalf of the society, for more information about his earlier successes. "[I heard that] you had at last . . . successfully accomplished that most difficult experiment on the transference of blood from one to

FIGURE 7: The first animal transfusion experiments were performed on dogs and usually connected the carotid artery to the jugular vein or the carotid artery to the femoral vein. Johann Sigismund Elsholtz, *Clysmatica nova* (1667).

the other of a pair of dogs." The "celebrated assembly" wished, Boyle explained, "for a more careful account of the success of the experiment." In a last effort to persuade Lower to lay bare his secrets, Boyle closed the letter with friendly praise: "There are many among its members who esteem you at your right worth and [who] are your friends, but none more so than Yours Affectionately, Rob. Boyle."[16]

Lower was more than happy to cooperate. His fingers black with ink, he spared no detail as he recounted the canine experiment at length, filling page after manuscript page about his choice of animals, their positioning on the table, the tools he used, the amount of blood that was shed, and of course the meticulous craft required. Lower made it clear to Boyle that he had every intention of continuing his bloody experiments; of that there was no doubt. But he was also eager to see just how far his model might be taken, offering detailed suggestions for those who might wish to try the procedure themselves. He had learned, Lower explained, that a small, weak dog did not make the ideal donor. Its heart beat too faintly, too quickly. "To prevent this trouble, and [to] make the experiment certain," he wrote, "you must bleed a great dog into a little one, or a mastif into a curr."

The surgeon's ongoing frustration, though, remained in the tools he used. Plucked from geese, quills were most commonly used for letter writing, not history-making surgical experiments. They were too small, and their quality—prepared as they were by decidedly unscientific quill dressers—was not always consistent. Instead of quills, then, Lower suggested that a specially designed "small crooked pipe of silver or brass" be attached to the end of each quill, serving as something of an anchor to a longer, man-made tube. Lower concluded his letter to Boyle by expressing his hopes that his trials would be "prosecuted to the utmost variety the subject will bear."[17]

Despite Lower's resolve to continue his work, blood transfusion trials would suddenly come to a halt once again when death refused to relinquish its hold on the weakened city. The last shovelfuls of dirt had barely settled on the city's mass plague graves when a fire, lit in a tiny bakery in Pudding Lane, devoured the heart of the capital on September 2, 1666. Narrow streets, timbered houses, summer drought, and strong eastward winds set ideal conditions for the inferno. Spooked and whinnying horses pulled carts loaded high with desperate refugees and whatever personal belongings they had been able to gather. Clogged streets were brought to a chaotic standstill in the frantic exodus from the city. Some lucky souls were able to make their way to the Thames and onto the overloaded boats that filled the river. The passengers set off directionless, torn between keeping an eye on the red glow of their city and the slowly rising water at their feet. Those who decided not to

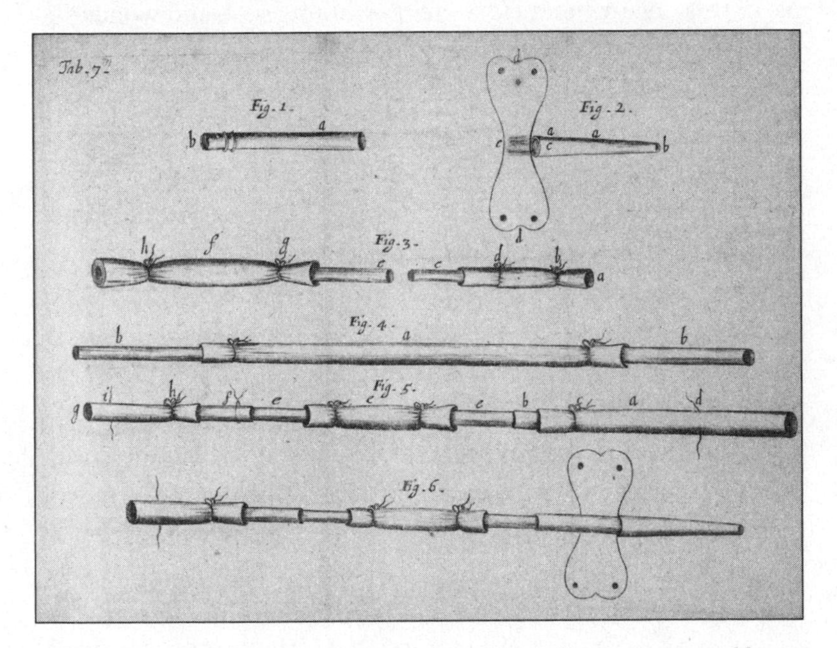

FIGURE 8: Richard Lower proposed a system of interconnected brass tubes that could be used in conjunction with quills for his early blood transfusions. *Tractatus de corde* (1669).

brave the frenzied streets remained in their homes, waiting and praying.[18]

The only defense against the searing-hot beast was to admit defeat. King Charles II ordered the lord mayor of London to "spare no houses and pull down the fire every way."[19] Perfectly intact homes exploded with the help of gunpowder; men with axes and shovels followed behind to clear a protective ring against the conflagration. The preemptive destruction helped to contain the fire, but "containment" was a relative term when more than 13,200 homes and 87 churches had already been incinerated. When the four-day inferno was finally extinguished on September 5, 1666, 85 percent of the buildings within the city's walls were little more than smoldering ashes.[20] Upwards of 65,000 people in this once-vibrant city were now homeless—and wondering when God's fury would end.[21]

To make matters worse England was deep in the middle of war. The two-year war between England and Holland that was thought to have been resolved in 1654 warmed up again in February 1665, just months before the plague. The Dutch were joined by the French eleven months later, in January 1666, and there were few signs that a truce was on the horizon. The French king Louis XIV was pleased about the fire's obvious strategic advantage for his country against its enemy. He did, however, prohibit any public rejoicing following the catastrophe and sent word of his willingness to lend aid to Londoners in their time of need. The gesture was greeted with appropriate diplomatic protocol, but there was little chance that the French king's largesse would be accepted.

In the days following the London fire, angry mobs needed to fix blame somewhere, anywhere. They set their sights on Dutch and French Protestants, who had come to England seeking asylum from religious persecution. Rumors spread as fast as the fire itself that a crazed Dutchman, assisted by the French, had delighted in throwing fireballs into the homes of innocents. The Londoner William Taswell detailed the random acts of violence that filled the streets in the days and weeks after the Great Fire:

> A blacksmith in my presence, meeting an innocent Frenchman walking along the street, felled him instantly to the ground with an iron bar. I could not help seeing the innocent blood of this exotic flowing in a plentiful stream down to his ankles. In another place, I saw the incensed populace divesting a French painter of all the goods he had in his shop, and, after having helped him off with many other things, leveling his house. . . . My brother told me he saw a Frenchman almost dismembered in Moorfields because he carried balls of fire in a chest with him, when in truth they were only tennis balls.[22]

One Frenchman—reportedly feeble-minded—nonetheless met his fate at the hands of the hangman after he confessed, erroneously and under coercion, to having started the fire in Pudding Lane. The anger of the British against the French was palpable in the shouts and taunts directed at Robert Hubert as he was led through the streets on his way to the gallows late that October. When an anatomist stepped forward to claim Hubert's body, he was shoved aside by the rabid crowd. By the time the masses were done beating and stripping the corpse, there was nothing left to anatomize.[23]

Three weeks after the fire the House of Commons appointed a committee of seventy persons to investigate its causes. The committee heard story upon story of "eye-witness" reports that "published with great assurance, came to nothing upon a strict examination." The committee's report was inconclusive regarding foreign involvement in the inferno, and Parliament had no choice but to drop the matter. Charles II and his ministers went even further and declared that "nothing hath been found to argue the fire in London [was] caused by other than the hand of God, a great wind, and a very dry season."[24]

Despite the horrific losses the destruction of London was for some a catastrophic—but decidedly welcome—prelude to a glorious rebirth. "Never a calamity, and such a one, was so well bourne as this is," wrote one observer just a week after the fire. "'Tis incredible, how little the sufferers, though great ones, doe complain of the Losses. I was yesterday in many meetings of the principal Citizens, whose houses are laid in ashes, who instead of complaining, discoursed almost of nothing, but a survey of London and a dessein for rebuilding."[25]

Once fascinated by blood circulation and infusion, Christopher Wren scrambled to draw up a new street plan for London over the six days immediately following the fire. Wren's ideas, which he presented to the king and his council on September 11, were clearly influenced by the young man's more recent experiences on the Continent. Now a member of the Royal Society, Wren had only just returned from a memorable trip to Paris. Planned well in advance, Wren's trip to the French capital in the plague-ridden summer of 1665 could not have been better timed. Instead of coming face-to-face with the hideous illness that fell other Londoners, Wren avoided contagion and explored the ever-growing marvels of the city in style.

The English and French capitals were a study in contrasts. While London struggled to survive, Paris nurtured its reputation as the seventeenth century's center of high taste and cutting-edge aesthetics. Wren had been sent to the French capital by Sir John Denham, Surveyor of the King's Works, to observe the flurry of construction going on there as part of Louis XIV's propagandist efforts to create buildings that were as impressive as the young king intended his reign to be.[26] Hammers and chisels echoed in the Paris air as the monuments that defined the Sun King's reign rose from the ground. Inspired by Saint Peter's, the Church of Val-de-Grâce was nearing completion, a first important step toward what Louis XIV's prime minister, Jean-Baptiste Colbert, envisioned for Paris as a "new Rome." Across the river from the Louvre, premier architect Louis Le Vau had been hard at work on the sprawling Collège des Quatre-Nations, with its equally soaring high dome and colonnaded facade. While major additions to the Louvre—home to French monarchs since the Middle Ages—were under way, work had also commenced on the remote outskirts of Paris, in Versailles, on a new royal palace. What was once a modest hunting lodge was set for eye-popping grandeur under the expert eye of Le Vau, the painter Charles Le Brun, and the landscape architect André Le Nôtre.

Inspired by what he had just seen in Paris, Wren eagerly got to work and began mapping a new London. While the influences of the neoclassical Italian and French models he had studied on the Continent figured prominently in Wren's ambitious design, his ideas for London's rebuilding also reflected the scientific and cultural attachment to ideas of circulation—both as medical model and metaphor—in Restoration England. True to his earlier medical inclinations, Wren began his work with a diagnosis. "Having shown in part the deplorable condition of our patient," he wrote in reference to the city of London, "we are to consult of the cure. . . . And herein we must imitate the physician, who when he finds a total decay of nature, bends his skill to a palliative to give respite for a better settlement of the estate of the patient." Wren further justified his plans by evoking one of the most central features of early medicine: the humors. It would be a "shame to the nation" and a sign of "the ill and untractable Humours of this Age," he argued, if London were to be rebuilt on its old foundations.[27]

Such a comparison between the health, humors, and the physical state of a city and its inhabitants was not at all novel in early Europe. As early as the first century, the celebrated Roman architect Vitruvius emphasized that there was a symbiosis between the human body and ideal architectural forms. In fact the human body was understood as being so perfectly symmetrical and proportionate that it served as a model for classical construction. Leonardo da Vinci himself celebrated the perfection of the human form and the centrality of man within the cosmos in his archetypal drawing of the perfectly proportioned and geometrically balanced "Vitruvian Man."

Yet for all that had been discovered about the body, there still remained so much to be learned. Natural philosophers and physicians yearned to control the strange and still unknowable mysteries of human physiology. In their attempts to make sense of the inner workings of the human body, they turned to external, man-made creations onto which they could project their developing physiological theories. Natural philosophers tried their hands at constructing animated, soulless life through machines—both human and animal.[28] The *Journal des sçavans*, the primary French scientific periodical of the late seventeenth century, included reports of physicians who "composed statues that resembled so well a man in all of its internal parts, with the exception of the rational soul, that it was possible to see everything that took place in our bodies, and this according to the principles of physics."[29] Their explanations were complicated, but the materials they used to produce these mechanical men were not: Fireplace bellows stood in for lungs, glass jars for skulls, corn grinders for stomachs. Still, the creators of such mechanical men could be true sticklers for detail. One writer

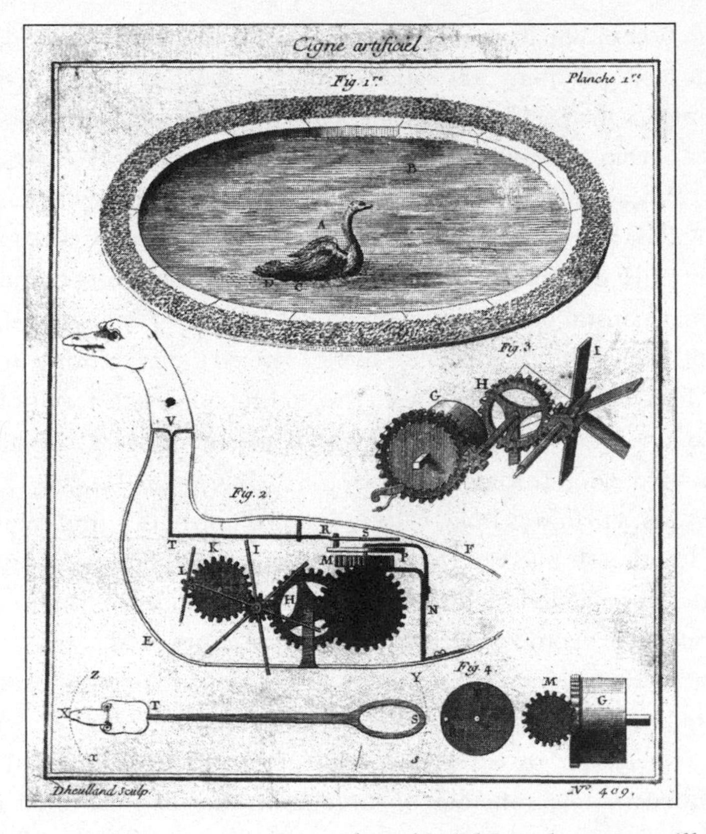

FIGURE 9: Late-seventeenth- and early-eighteenth-century illustrations—and the machines that were built using the drawings—demonstrated the mechanical processes underlying an animal's movement. Here, the swan's paddling through water is shown in terms of cogs and wheels. "Diverses machines inventées par M. Maillard. Cygne artificiel" (1733).

noted that the male "gland [penis] needs to get larger and lengthen," therefore "some wrinkled and soft [animal] skin" must be used "so it can dilate and shrink easily."[30]

The seventeenth century ushered in an age of machines—and circulation stood as a prime exemplar of mechanistic understanding. When William Harvey plotted the circular path of blood, he was also describing an elaborate human machine composed of pumps, valves, and tubes. And it was through circulation—both as a metaphor and a scalable model—that Wren had expressed his radical new vision for the city of London following the Great Fire. For Wren a new layout for the city would not only cure London's ills: The plan was the ultimate physical representation of a mechanistic human body and a celebration of the circulation models that the architect had spent so much time exploring in his early career at Oxford. The city would function like a well-oiled machine, moving people and goods as smoothly as the heart pumped blood through the body.

Wren's plan was bold, calling for a complete raze and rebuild of London. It rejected the tight network of serpentine medieval streets as plague-breeding firetraps. Instead wide thoroughfares would cut geometric angles through the city, and open piazzas would ensure a constant ebb and flow of the city's crush of cart and carriage traffic. The design called for a free-flowing city made up of arterylike roads and circular piazzas that echoed the very themes of circulation that had captured the scientific imagination of seventeenth-century England. Two wide avenues emanated from the east. Serving as major arteries, one thoroughfare flowed from the Royal Exchange, the other from the Tower of London. The two would come together at Wren's rebuilt Saint Paul's Cathedral, which would serve as the heart of the city. United at the cathedral, the avenues would continue across a now-cleaned-up Fleet river until they came to a "head" at an enormous circular piazza. London, in Wren's plans, "would be like a human body in which the unhindered circulation of people, money, and goods would nourish and bring to life all corners of the capital."[31]

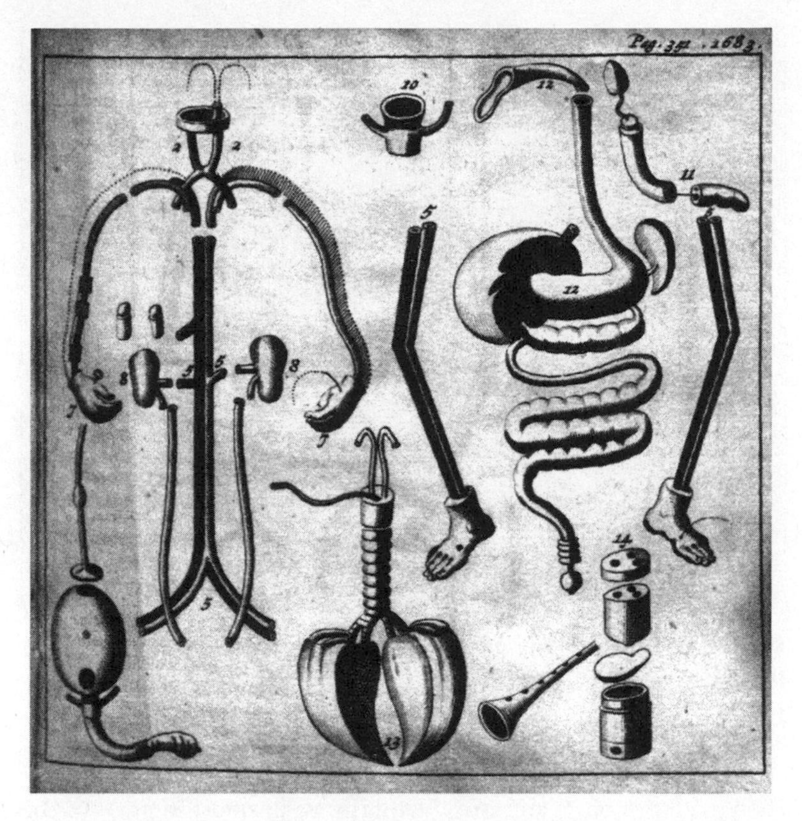

FIGURE 10: The Statua Humana was constructed from simple objects such as funnels, tubes, and bellows. Image nine illustrates the flexible male member, which is enlarged by pushing on the balloon-like bladder sitting directly above.

Yet postfire construction efforts presented a frustrating reminder that politics were not subject to the rules of anatomy and physiology. While well received, Wren's plan had not been the only design presented. Evelyn offered his own three days later, and then Robert Hooke added still another, which was itself followed by those of city surveyor Peter Mills, cartographer

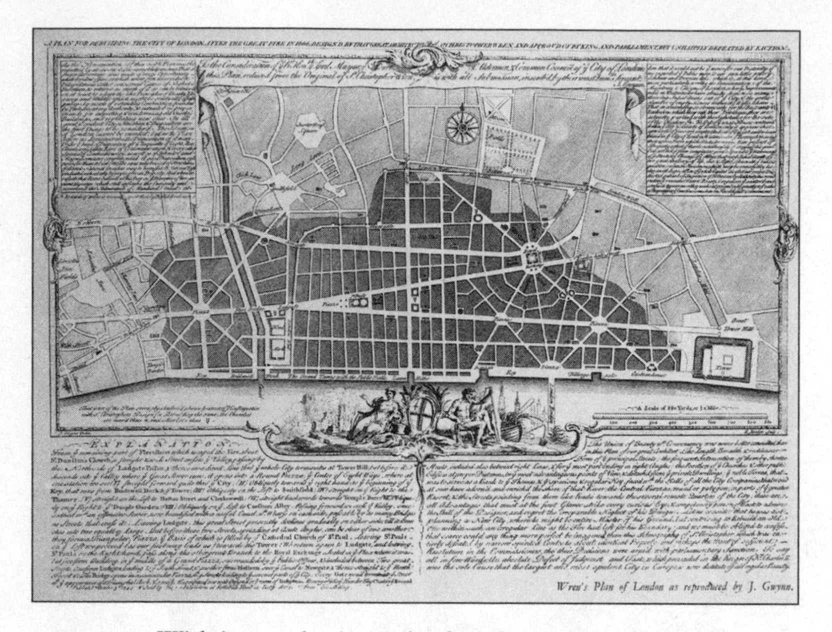

FIGURE 11: With its emphasis on the free flow of Londoners through the city's rebuilt "arteries," Christopher Wren's plans for postfire London reflected Harvey's circulation studies.

Richard Newcourt, and army captain Valentine Knight. As the merits of each plan were debated, the determination of property lines and the rightful ownership of plots proved more difficult than anticipated. Many property owners had fled the city and had yet to return to claim their now-destroyed homes. They also faced exorbitant surveying fees, which seemed to increase daily, as demand for surveyors far outstripped supply. And of course many property records had been destroyed in the conflagration, creating a legal mess that would have to work its way through the courts for years to come. In the end Wren's ambitious designs for London were shelved along with those of Evelyn, Hooke, and the others. London would be rebuilt on its old foundations.

Despite the eventual rejection of his glorious plans for London, Wren turned his efforts to a project that would take nearly thirty years to complete and one that would be the crowning glory of his architectural legacy: Saint Paul's Cathedral. However, other members of the Royal Society, like the steel-willed Richard Lower and his colleagues Edmund King and Thomas Coxe, returned as soon as they could to the laboratory to begin anew their blood experiments and to push their trials to the logical end point: human transfusions.

Chapter 5

PHILOSOPHICAL TRANSACTIONS

So it was on the evening of November 14, 1666, that candles and lamps filled the Royal Society's long, narrow meeting room, casting shadows on several dozen well-dressed men as they nudged closer for a better view.[1] Wooden floors creaked and conversations echoed off the imposing thirteen-foot ceilings. Two eager surgeons, Edmund King and Thomas Coxe, had been given the honor of replicating Lower's groundbreaking experiments. They knew that they had very large shoes to fill, and they moved busily through the room as they prepared the table for their victims.

In the effort to re-create Lower's earlier canine transfusion experiments to the letter, two dogs were brought in: a spaniel and a mastiff, which were lifted up to the experiment table and tightly secured. The room was vibrating with anticipation as eager spectators elbowed their way closer and closer to the table. Soon the din was punctuated with perfectly orchestrated, high-pitched yelps as the surgeons' blades sliced first through the vein of the spaniel and then through the artery of the mastiff. The floors were slick with blood; the surgeons' clothes were stained red. Like a triumphant bullfighter, one of the surgeons ordered the now-dead mastiff to be carried away from the table, out of the ring. Glad to be released from its bonds, the spaniel jumped spiritedly off the table and ran to the door as fast as it could. In the days and weeks that followed, the spaniel grew—not as big as the mastiff had been—but he still grew and got fat to boot. It had been, as Pepys later described it, a "pretty experiment."

Towering in the distance during the experiments was the stout Henry Oldenburg. As secretary and publicist for the Royal Society, he was responsible for setting his own inked quills into action and spread the news of yet another English success. If Oldenburg had risen quickly through the ranks of the tight-knit society, it had to do in no small part with his extraordinary linguistic abilities. German by birth, Oldenburg had spent his youth on the Continent and was fluent in English, French, Italian, Latin, Greek, and, of course, English. While the secretary of the Royal Society was exceptionally conversant with the science of the day, he was not a natural philosopher per se. His name had been built around his abilities as a polyglot reporter, and it was likely that his connections with Robert Boyle and other strategically placed correspondents had ensured his membership in the society. Oldenburg was among the first forty men to be nominated to the society and was officially elected to its ranks on December 26, 1660. He was later named as one of two secretaries, along with John Wilkins—a position he would hold until his death in 1677.

Oldenburg clearly took his responsibilities as secretary very seriously. When the society's weekly meetings were discontinued during the plague, he was one of the few members who did not flee to the countryside. Remaining in his home in Pall Mall, he continued his correspondence with colleagues in England and abroad—no easy task, given the substantial disruption of courier services. Oldenburg was so worried about the "books and papers belonging to the Society that are in my custody" that he had made contingency plans for these materials to be put in a box and transferred into safe hands should he feel unwell.[2]

Five days following King and Coxe's blood transfusion at the Royal Society, Oldenburg closed his *Philosophical Transactions* with a short, and much understated, article about the experiment:

> *The Success of the Experiment of Transfusing the Bloud of One Animal into Another*
>
> This Experiment, hitherto look'd upon to be of an almost unsurmountable difficulty, hath been of late very successfully perform'd not only at Oxford, by the directions of that expert Anatomist Dr. Lower, but also in London, by order of the R. Society, at their publick meeting in Gresham Colledge: the Description of the particulars whereof, and the Method of the Operation is referred to the next Opportunity.[3]

With satisfaction Oldenburg returned his quill to the inkpot and reached for the lead pounce pot that sat close by on the writing desk. The ink was still wet; a sprinkling of fine pounce powder would help dry it quickly. He collected the manuscript pages, skimmed through them attentively, and then sent them without delay to the printers in what was then Duck Lane in the vicinity of Saint Bartholomew-the-Great.

Duck Lane was a narrow thoroughfare that sat in a northern London neighborhood called Little Britain. Just one street outside of the perimeter of the fire's destruction, Little Britain had quickly become the makeshift headquarters for publishing. The stationers—as publishers were called—had seen their stock and storage go up in flames at Saint Paul's Cathedral. Even the books and wares they had carried for safety into Saint Faith's Church lay now in ashes, along with the latest issue of the *Philosophical Transactions.* The printers were unsettled and broke; what little printing equipment and paper remained had become extraordinarily precious—and expensive. Oldenburg had to do no small amount of negotiating to ensure that a new issue would be produced.

With October's cinders still smoldering, it was now mid-November and the weather in London was cooling with each passing day. But here in the small space of the noisy and bustling printer's shop, there was no chance of escaping the heat. Metal clanged, printing presses rattled. The cacophony was made still louder by the anxious shouts and name-calling that broke out regularly in the workshop. Publishing took laborious effort and required meticulous attention to detail. One mistake could give a commissioning author reason to ask for reparations or, worse, demand a new print run.

In the corner stood wooden cases filled with letters, punctuation marks, and spaces. A typesetter gathered letters with lightning dexterity as his assistant read aloud each word of Oldenburg's manuscript. Once the typecase was full, the typesetter gestured to an apprentice, who gingerly carried the tray to an enormous table where it waited to join others for its turn at the press. In the center of the room the pressman rolled oil-based black ink evenly across the raised letters with a ball. Sleeves rolled up and sweat dripping from his brow, he expertly squared a large folio of damp rag paper on the cases, reached up for the handle, and lowered the press onto the pages. Bearing down for two or three seconds, he then freed the paper from the press, gently peeled the page from the form, and hung it to allow the ink to set and the paper to dry.

Page after page, hour after hour, the process repeated itself. The pages were folded and sewed together. A few days later nearly twelve hundred copies of the *Philosophical Transactions* would be on their way to all corners of London, its surrounding countryside, and even farther.[4]

As they would for months and years to come, scientists and medical men scrambled to arrange for this latest issue of the *Philosophical Transactions* to be sent to them through an informal network of acquaintances. Couriers and "mercuries" pushed one another aside as they clamored for copies of the journal. Mostly women of various reputations, mercuries were resellers who "cried books" from their stalls at the London Exchange. The printers spied each woman as she entered the shop, and they agreed to sell their wares—still reluctantly at that—only to those whom they trusted. A good profit could be lost quickly to mercuries and other hawkers who regularly forged cheap copies of stationers' products.

While most copies of the *Philosophical Transactions* remained in England, a privileged handful would travel across the Channel and wind up in the private libraries of the highest-ranking noblemen. Once the precious goods were firmly in the hands of the noblemen's couriers, they moved east through the still-smoldering city by way of Watling Street, which led directly to Canterbury. From there the couriers made their way through the woods over the narrow and muddy road to Kent, and then forged ahead through the hills to Dover. From the cliffs of England's premier port city to the Continent, it was possible to spot the French town of Calais. The *Philosophical Transactions* would find land again there, where another courier waited to shepherd the precious publication on the final leg of its journey to Paris.

Oldenburg maintained close connections to France. Between 1659 and 1660, before his appointment to the Royal Society, he had spent several months in Paris as the tutor of Robert Boyle's nephew. Now, as London struggled to rebuild, it was Boyle who encouraged Oldenburg to stay in touch with acquaintances in France to keep an eye on the scientific activities of England's enemies. With the country in the grip of war, the Royal Society secretary forged, in particular, a strategic correspondence with Louis XIV's secretary, Henri Justel, who offered an abundance of political updates as well as juicy gossip about life among the French nobles.[5] Oldenburg had learned confidentially from Justel that plans were afoot to form a French Academy of Sciences. "Steps are being taken here," Justel wrote, "towards the establishment of some academy to be composed of men selected from all sorts of professions. We do not yet know the details of it, for that is only sketched. If the idea is taken to heart, some considerable establishment will result, and there is reason to hope it will succeed. Do not speak of it very definitely until it is further advanced."[6]

By the end of January 1667 rumors of English experiments with blood transfusion were buzzing quietly throughout medical Paris. Oldenburg's short announcement had been supplemented a month later by Lower's complete letter to Boyle in the next issue of the *Philosophical Transactions*. But the few copies of the *Philosophical Transactions* that made it to Paris were held tightly by high-ranking nobles who had either the financial means to buy up the rare copies or could tap their personal connections to Oldenburg to acquire one. For men like Jean-Baptiste Denis who were not part of the moneyed scientific elite, their knowledge of the English experiments was based, then, on little more than hearsay and speculation. Yet even if Denis had been privileged enough to gain a rare peek at Oldenburg's publications, he—like most other men of science of the time—would not have been able to make heads or tails of it. The French medical community—even those who now had the news of English transfusion in their hands—would still have to wait for translations to be completed.[7]

The delay riled Denis; he was eager to stay on top of the most-cutting-edge medicine. His characteristic impatience later led the French physician to write directly to the multilingual Oldenburg and beg that French translations of the journal be made available. "I wish I understood English in order to be able to read your Transactions; whenever I can find someone who explains something to me I am even more eager to see the remainder. I have often wished that instead of the copies in English which you send to France you would send just one in French; I would gladly have had it printed at my expense both for your own reputation and the gratification of an infinite number of the inquisitive, who would be delighted to be able to read and understand them by themselves, instead of which there are only three or four who see them."[8] Living on what little income he earned giving anatomy lessons to beginning medical students, however, the ambitious Denis would have been unlikely to make good on his expensive offer. But that was soon to change.

Denis' modest apartment on the Quai des Grands-Augustins overlooked the gray waters of the Seine where it narrowed and wrapped around the Île de la Cité. The largest of Paris's several islands,[9] it served as a transition between the city's Left and Right banks. The Left Bank, where Denis lived, was home to the Latin Quarter. This vibrant area of Paris earned its name from Latin, the lingua franca used in the universities and in the bookshops that populated the quarter's streets and alleys. Serious students in their ankle-length black gowns memorized the works of ancient philosophers and studied the effects of cheap wine in their evening revelries. Just steps from Denis' home were streets like the narrow rue de Seine, where French students mingled with Flemish, Dutch, and German students who had come to France seeking adventure. And in the grungy boarding houses where they lived, rats, mice, and fleas kept all of them good company and "mistreated" them at night.[10]

Newly married,[11] Denis had only recently begun to settle into his adult Parisian life, living among the students in the Latin Quarter. Just a few months earlier he had been finishing his studies in Montpellier's renowned medical school. The son of an artisan, Denis knew that he would be fighting the odds in his attempts to establish his reputation among Paris physicians and their prominent patients. Yet he remained more confident than ever in his abilities to treat even the most stubborn of illnesses. As a young child he had suffered terribly from asthma. Despite countless remedies, family doctors and apothecaries had been unable to provide relief. But as Denis reports in one of his many treatises, he alone had been able to cure his ailments and been able to keep his asthma at bay through the help of self-prescribed treatments of inhaled sulfur.[12] The thirty-two-year-old eagerly looked forward to his next trip to Montpellier, when he would be awarded his doctoral "bonnet." The cap would signal to the world that he had been accepted fully into the highest ranks of medicine—and that he had transcended his bourgeois origins to enter, as only rare men did, the elite ranks of society.

Medical training in France, as throughout Europe, was steeped in traditions that had endured since the thirteenth century, when the first major teaching faculties had been founded. After completing their bachelor studies, students from well-placed families requested admission into the medical schools, where they attended lectures and suffered through oral exams before serving an apprenticeship. Like the courses at the university, the apprenticeship rarely focused on hands-on patient care. Lowly surgeons were the ones who worked directly with patients—and often in painful and unpleasant ways.

The physicians took a more philosophical approach to health. Their knowledge was squarely rooted in the texts of Galen, Hippocrates, Aristotle, and Avicenna, among others. The students learned about the structures and functions of the human body primarily through the droning words of their instructors, who recited, often verbatim, the works of these old masters.[13] And the students scribbled word for word their teachers' magisterial lectures with cramped and ink-stained hands. They did not have the luxury of thinking independently, much less forming their own thoughts, during the lessons.

As a Montpellier-trained physician, Denis underestimated how hard it would be to navigate the strict hierarchies and brutal politics of the Parisian medical world. Since at least the late sixteenth century, rivalries and tensions between the doctors of the craggy hills in the South of France and those who trained and practiced in cosmopolitan Paris had been intense. Montpellier produced nearly 40 percent of all physicians in France, but the university had a troubled reputation as a party school where medical students were just as likely to drink and cavort with prostitutes as they were to learn the intricacies of the Hippocratic corpus. As the philosopher Julien Offray de La Mettrie wrote in the following century: "In Montpellier, those who are destined for medical careers are for the most part young good-for-nothings who give themselves over to dissipation and indulgence for the first two years of their studies. It is only in the third that they even begin to study, in order that they may reply to frivolous questions such as: *Quid est vita?* [What is life?]"[14]

Disdain for Montpellier's medical school and its long-standing reputation for unruliness was linked, no doubt, to the fact that the school had long accepted Protestants among its students. The faculty had also shown itself open to pursuing new ideas, especially those of William Harvey—who was himself a Protestant.

Since as early as the 1650s, Harvey had supporters among the University of Montpellier's faculty members. And within the ten years that followed, circulation had established itself not only in the teaching curriculum there but also among the less-trained Montpellerian surgeons.[15]

This was not at all the case farther north at the University of Paris medical school, where Harvey's theories were still most unwelcome. Medical education and practice in France's capital had long been steeped in the traditions of the Paris Faculty of Medicine. The medical school prided itself in being among the last and most prominent strongholds against blood circulation. Nearly forty years after Harvey's *De motu cordis,* the "anticirculator" argument continued to lay a claim on the hearts and minds of students in the French medical school. In the early years following Harvey's claims, a student at the University of Paris named Simon Boullot defended a thesis that denounced circulation. Using a traditional Galenist argument, Boullot held firm that circular movement was too straightforward and was suitable only for the simplest of creatures. In humans blood crossed the chambers of the heart through invisible pores to be transformed "into a pure spirit and juice, which keeps warm the native heat of all of the members." Boullot argued strongly that there was nothing in the structure of veins and arteries—seen or unseen—to indicate that blood was pumped throughout the body by the heart.[16] He passed his thesis defense with flying colors and with the great praise of his examiners.

In 1645, however, it looked as if circulation might have a chance of success at the University of Paris. A student named Jean Maurin presented an ambitious thesis titled "Whether, because of the circulatory motion of the blood in the heart, Galen's method of healing is to be changed."[17] Maurin gave credence to the idea that blood vessels formed a branching and interconnected network (anastomosis) that could account for the circulatory path of the blood. The thesis was approved by his committee, largely because it was chaired by Jean Riolan, one of the few members of the faculty who had shown a quiet interest in circulation. In 1648 Riolan himself finally spoke out publicly in favor of circulation. Yet even he remained attached to basic Galenic notions that blood was produced by the liver. Riolan also argued that blood took a full two or three days to circulate completely through the body.

This and other "circulator" arguments were patiently tolerated by the rest of the Parisian medical faculty, but it soon became clear that tolerance would be the best, if not the only, thing that circulation's proponents could hope for. In later years the dean of the Paris medical school would sign off on two student theses that condemned circulation theory: *An sanguis per omnes corporis cenas et arterias jugiter circumferatur? Neg.* (Is the Blood Conveyed Continuously Through All of the Body's Veins and Arteries? Negative), and another with a circular title of its own, *Est-ne sanguinis motus circularis impossibilis? Aff.* (Is Not the Circular Motion of Blood Impossible? Affirmative).[18]

Filled with naive hubris, Denis came to Paris from Montpellier ready to make a name for himself in a city where he was clearly at a disadvantage, both by birth and by training. Denis began his efforts to make connections with the Paris medical community by offering anatomy tutorials, for a fee, to medical students and other curiosity seekers. When human dissections were first performed the late Middle Ages, physicians sat in elevated chairs above both the corpse and the barber-surgeon. Book in hand, a physician-professor explained basic anatomy and gave orders to a surgeon to reveal the specific body part under discussion. Space was too tight in the small rooms of Denis' home for such a magisterial design. A small platform alongside the dining room table on which the cadaver was displayed was likely all that was possible. Yet this did not keep Denis from projecting the overconfident demeanor of a newly minted physician. Following the more hands-on approach of the Renaissance anatomist Vesalius, Denis performed portions of the dissections himself. With a large leather-bound anatomy sitting alongside the corpse, Denis took apart the cadaver—layer by layer, piece by piece—while young medical students leaned in as closely as they could, even at the risk of being overcome by its rotting stink. (We have no record of what the newly married Mrs. Denis thought of her husband's craft in their home. It is probably safe to say she was not pleased.)

Denis continued to wait eagerly for more detailed news of the English successes in transfusion. After nearly two months' anticipation, the French *Journal des sçavans* (*Journal of the Erudite*) finally printed a translation of Oldenburg's announcement regarding Boyle's presentation of Lower's experiments to the Royal Society. On the last day of January customers spilled onto the cobblestone street in front of Jacques Cusson's printshop on the rue Saint-Jacques. Denis was no doubt among the greedy crowd who clamored for copies of the paper. A few flips of the pamphlet pages, and there it finally was—a faithful rendering in French of England's claims on canine blood transfusion.[19]

The young doctor devoured the details and set to work to try his own hand at the procedure. His first order of business was to review, in the presence of his paying students, the anatomy of the cardiovascular system. Assisting Denis, the surgeon Emmerez rolled his sleeves up above the elbow and fastened a rust-stained apron around his waist. The cadaver had been gutted nearly as soon it had been brought into Denis' home, the entrails tossed into the nearby river; the softest organs were always the first to rot. Emmerez reached into the open cavity, pushing and probing body parts as Denis narrated the path of blood that had once flowed through the man's dead body. Denis recited Harvey's main propositions regarding circulation. First, the production of blood began in the digestive system. Second, it flowed to the heart from the inferior vena cava with the help of an elaborate system of valves that ensured the one-way direction of its flow. To demonstrate the point Denis sliced open a large vein; he pushed the tip of his knife through the valve in one direction with relative ease. Then coming at it in the opposite direction, he showed how the closed venous valve blocked his tool. The men then turned their attention to the heart, which they removed from the body and placed on a nearby table, and explored the chambers.

Denis' demonstration was little more than a recitation, with visual aids, of what had become conventional wisdom over the three decades following William Harvey's groundbreaking work. However, the French doctor could not contain his excitement about what he had been hearing regarding transfusions by the English. During one of his dissection demonstrations, he shared his conviction that transfusion was "new and completely convincing proof" of the truth of circulation. But instead of respectful nods, he received chortles and chuckles from his audience, which was populated largely by students from the conservative University of Paris medical school. Blood transfusion, they retorted, was too "chimeric and ridiculous" to be believed.[20] Furious, Denis ordered the incredulous audience out of his home. He believed, like every highly trained physician in the early modern era, that his words demanded respect. He was seething.

Not one to take humiliation lightly, Denis would do everything possible to prove them all wrong. A few weeks later he once again enlisted the help of the surgeon Emmerez to experiment firsthand with blood transfusion. They arranged to have two small dogs brought to them. One was a tall, fat female spaniel, and the other a small, skinny, short-haired male that looked something like a fox. The dogs, Denis later explained, "had never been fed together and their appearance was so different that they seemed almost as if they were animals of different species." Denis resolved not only to replicate the English experiments but also to go a step further. He would transfuse blood into one without killing the other.

In front of a group of carefully selected supporters, Denis and Emmerez embarked on their trial. They began by muzzling the dogs to "keep them from crying"[21] and then positioned the animals head to foot, so that the thigh of the recipient almost touched the neck of the donor. This was a two-man job. Scalpel in hand, the doctor and his assistant followed Lower's well-choreographed procedure to the letter. They marveled as arterial blood pulsed through the system of small tubes that connected the two animals. Another tube emptied the recipient's own blood into a shallow bowl. Though it seemed to Denis that the amount of blood flowing into the recipient equaled the blood coming in, there was no way to know for sure without checking. From time to time Denis and Emmerez gingerly disconnected the two tubes and confirmed that blood was indeed moving through them. And they noted with pleasure that it was gushing too fast and remained too hot to clot. They delighted as well at the regular pulsations they could feel in the throbbing vein of the recipient. The platter continued to fill. Nine ounces had been let from the recipient and, presumably, were replaced by another nine ounces from the donor.

Then, suddenly, the experiment took an unwelcome turn. The spaniel weakened measurably and looked very close to death. Without hesitating Denis ordered Emmerez to stop the experiment and begin stitching up the dogs. The spaniel who had unwillingly donated blood remained weak. It had only enough energy to crumple into a corner of the room. The other dog was "vigorous" and attempted to scratch off its muzzle. It jumped down from the table, shook itself, and stumbled over to its owner for treats and pats when called. Denis admitted, however, that the dog was obviously not quite as "awake and gay" as before the trial. It had been a painful and exhausting procedure for both animals. To be sure that the animal's sluggishness was a matter of the discomfort of the incision rather than an effect of the transfusion itself, Denis performed a control experiment. He brought in a third dog of similar size and made an identical incision in the jugular vein. Once stitched back up, this dog appeared even more "beaten" than the one that had been transfused. And though the transfused dogs ate heartily in the two hours following the experiment, the third subject refused to eat.

Denis kept all three dogs in his small apartment during the week that followed. He noted their every movement, every morsel they ate, and compared their weights. Denis may not have mentioned the disarray that the animals must have caused in his home—or his wife's displeasure of having unruly canine houseguests—but he did gloat with pride that all three dogs returned to perfect, and playful, health in short time. He also noticed an odd side effect of the procedure. It turned out that the donor spaniel had been pregnant. The dog miscarried a few days later and, curiously, Denis reported, "only three or four drops of blood could be found" in the offspring.[22]

One week later a new experiment was in the works. There was no need to confirm the utility of blood transfusion; Denis took that now as a given. Instead, this next experiment would test orders of magnitude. If one dog's blood could be transfused into another, would it be possible to exchange blood among three different ones? On March 8, 1667, Emmerez set out his surgical tools once again on a makeshift operating table in Denis' dining room. It had not been easy to herd three rambunctious dogs up onto the table, especially when they already knew what was in store. But with the help of tight ropes and muzzles, they had been restrained.

Denis and Emmerez positioned the three dogs head to toe and toe to head. Preparing for what would be a round-robin of transfusions, the two men focused their efforts first on the spaniel and the foxlike mutt, their original donor and recipient. Now they bled the mutt into the spaniel, taking the donor to the point of near-death. Emmerez deftly stitched up the spaniel and released it from the table. Moving quickly, Denis turned his scalpel onto the third dog, whose blood would be used to reanimate the mutt.

The room was getting cold, and the two men yelled out to the handful of viewers to stoke the fire—and fast. A cold room would hasten the dogs' deaths and was also causing the blood to clot in the transfusion tubes, which were longer than the ones they had used before. A raging fire now crackling in the background, the third dog was bled into the previous donor. At frequent intervals, the men disconnected the tubes, warmed them with their hands, and blew forcefully into them in a frantic attempt to dislodge the blood that was now clotting. Nearly twelve and a half ounces of blood had been emptied from the mutt, and now sat in the shallow dish nearby. Denis knew better than to say that they had been able to replace all of this with blood from the third dog. But what he could say for sure is that at least some of the blood made it in the mutt.

When the tense experiment was over, the dogs each skulked into the corner, where they whimpered in pain and distress. Denis had planned to let the animals quietly recuperate before he performed follow-up tests of their appetite, weight, and stamina. However, he soon discovered that a spectator had slipped at least one of the dogs a large gulp of wine. This went a long way to explain why the animal was lurching like a drunken sailor as it walked. Annoyed as he might have been by a meddler in his work, Denis was nevertheless delighted with the knowledge that his procedure had been a success. All three dogs had survived—just barely, but they had survived.

While his initial experiments had been motivated by his desire to prove naysayers wrong, his interests in blood research had quickly morphed into a selfish recognition that transfusion could very well be a way to catapult him into celebrity. In his March 9 report to the *Journal des sçavans* he announced that he would soon be taking his show on the road. He issued an invitation to one and all to join him on the banks of the Seine at 2:00 p.m. the following Saturday to witness another, even more amazing blood transfusion: "We propose now to give you public proof. So you can see what changes transfusion can produce, we will transfer the blood of a young and healthy dog into the veins of an old and mangy one."[23]

Now, at the foot of the legendary Pont-Neuf, a noisy crowd of observers lined the banks of the river. A good dissection or public surgery always brought out the city's finest: amateur scientists, noblewomen, street children, beggars, and thieves. Noblemen in powdered wigs also dotted the chaotic landscape; they could be spotted in an instant by the perfumed white handkerchiefs they held to their faces to ward off the stench of the unwashed masses. A hush fell over the crowd as Denis stepped center stage. Everyone listened intently to the transfusionist's praises of blood's mysteries and how he and he alone had mastered its secrets. Nodding to Emmerez, Denis stepped forward solemnly to begin the transfusion. Under winter clouds the two men made good on the promise to transfuse life into the elderly dog. Again both animals survived. Denis had just publicly established his reputation as the premier transfusionist of France.

Henri-Louis Habert de Montmor was among the noblemen watching Denis' circuslike show with a mix of curiosity and excitement. Montmor was neither a physician nor a natural philosopher himself. While he was well aware of the tensions between Montpellier and Paris, his focus was trained squarely on his own self-interest. He still held fast to a quixotic dream of leading Europe's most influential private scientific community—an academy that would rival England's Royal Society and, especially, King Louis XIV's nascent Academy of Sciences. Like Denis, whose initial blood transfusion experiments had been fueled by perceived insults, Montmor was himself smarting from what had been his unceremonious dethroning as private benefactor to the sciences.

In France science had long been supported by a fragmented system of private patronage, with less than stellar results. Wealthy men like Montmor competed for the period's most celebrated thinkers in order to solidify their own social standing in Paris. Yet the French had long lagged behind the English when it came to scientific research. And Royal Society members like Oldenburg had not been shy to acknowledge this. "One must admit," he bragged, "that the English surpass [the French] and have the advantage over the other peoples of Europe, for they have given us a quantity of curious facts, in addition to the great books which they have published. On the contrary, the books published in Paris do not deserve to be read, at least most of them, being nothing but reiterations and assertions; without the facts which satisfy the mind."[24]

But just as the young Louis XIV and his indefatigable prime

minister, Colbert, had used the buildings of Paris and Versailles to establish the unquestioned grandeur of the French monarchy, they would soon expand their focus to building a glorious royal regime on the foundations of state-sponsored science. The French Royal Academy of Sciences was only a few months old when Denis had taken to the streets to put his blood transfusion show on display. The unknown transfusionist from Montpellier was, of course, not part of the elite handful of men who had been appointed to the king's academy. Nor was the now-displaced Montmor. And as Montmor watched Denis' every move that day on the Quai des Grands-Augustins, he knew that he had found the man who would help him restore his private academy to its former glory. Together he and Denis would take on not just the English but the king of France as well.

Chapter 6

NOBLE AMBITIONS

Like most rich Parisian noblemen who fancied themselves amateur scientists, Henri-Louis Montmor had long made it his routine to visit the artisans who sold their pricey wares on the island's Quai de l'Horloge.[1] Sitting squarely between the Left and Right banks on the Île de la Cité, the Quai de l'Horloge was a premier address for those in search of the best that Paris had to offer: rare gemstones, oil paintings by the masters, collector's coins made of the purest metals. But during the regular trips that he took to the quai, Montmor always had a singular goal: to bring back expertly crafted instruments and other rare curiosities. He put these tools and toys proudly on display for the many guests who streamed into his stately home on the Right Bank. The shelves of his library creaked with the weight of brass globes etched with the most recent cartographic discoveries, small microscopes embellished with the finest artistic designs, pea-size lodestones that attracted objects nearly one hundred times their size, and the very latest novelty to capture the fancy of well-heeled Frenchmen: barometers that could measure the very weight of air.[2]

The small shops nestled under the turreted towers of the Quai de l'Horloge bustled throughout the day with well-dressed shoppers, both French and foreign. While visitors often complained of the strong fish smell wafting from the boats of the carp sellers below, their displeasure quickly dissipated when they were welcomed into the workshops of the artisans whose craftsmanship had earned them the rare title of "engineers."[3] Like bookstores and, much later, cafés, instrument makers' shops were privileged spaces where nobles could meet, socialize, and marvel at the ingenuity of French craftsmanship.[4]

Shopkeepers offered a wide array of quadrants, foldable rulers, protractors, and compasses at all price points. But large sections of most stores on the Quai de l'Horloge were dedicated to sundials, which had become a visible and required mark of nobility in this last half of the seventeenth century. Clocks had not put sundials out of business, nor were they likely to anytime soon. A well-made clock was far beyond the means of even some of the richest families, and mechanical timepieces were notoriously inaccurate, measuring time only in one-hour increments.[5] The art of dialing, as it was called, was not only important for setting clocks straight—it had also become an integral part of polite culture, a mark of good breeding and high status.[6]

Like most men Montmor was drawn to elaborately engraved sundials that folded up and fitted in a small box that was itself stored in another protective case. At about 2.5 inches in diameter, the whole package could fit nicely in his coat pocket.[7] He was also one of those discriminating customers who could demand such *premiers cadrans* (top-of-the-line sundials), made of silver, without flinching for one moment at the price. Indeed, to avoid the indiscreet topic of money, artisans probed new customers about their preferences: silver, brass, or ivory. The ultimate choice of materials revealed everything worth knowing—and the level of service customers were likely to receive. As disappointing as it was, artisans also knew that the most elaborate and unusual scientific pieces commissioned by many of their customers—from pillar sundials embedded in the heads of walking sticks to sundials that doubled as pocket knives—were destined to be only exquisite curiosities. It was part of a required show that confirmed both intelligence and wealth among nobles.[8]

Among the high-ranking French of the seventeenth century, science was more of a spectacle and a show of social status than anything else. Pocket-size sundials allowed men like Montmor a portable means to display their wealth and their presumed learning. This consumerist fascination with the sun, the stars, and the heavens was exhibited as well by the imposing telescopes that jutted out like "deadly weapons from the roofs of peaceful citizens" all over the wealthy quarter where Montmor lived, the Marais.[9] A fascination—a fetish, really—for telescopes had taken over the French capital. A yearning for the stars had preoccupied all of well-heeled Paris, from the ladies in the salons to the university men. As with most everything in this ostentatiously elite society, the telescopes had been mounted at great expense as a display of access to knowledge—and, like so many other tools of science at the time, a fashion statement.

Observation parties established themselves as a regular feature of social life. The telescopes extended more than twenty-five feet and were hoisted into the skies at steep angles with the help of a large post that looked something like a ship's mast. The largest telescopes required support from a triangular joist attached to the instrument's midsection, to help prevent sagging in the middle. Stunning pieces of craftsmanship, telescope exteriors were made of exquisite leather and metal appliqué. But inside they were little more than a tube of parchment or thick cardboard that connected an eyepiece to a rudimentary lens. Once secured to the mast, the instrument was threaded into a decoratively embellished stand. And on clear nights a comfortable and equally elaborate chair would be brought to the roof for excited turn taking among nobles, who could be counted on to confuse the North Star with Venus or even with the moon itself.

Montmor collected men as greedily as he collected these and other scientific playthings. In the years preceding the establishment of the French Academy of Sciences, the nobleman had opened up his affluent home to the most prominent thinkers, explorers, and social brokers of his time. Montmor's dedication to science—and his affluent pride—were forged into the very stones of his residence and were made visible to all inquisitive outsiders who were willing to risk life and limb to peer inside the nobleman's compound. The entryway was two stories high, rather than the standard three to four, and was flanked by two short walls, just over ten feet at the highest point on each side. For those adept at dodging carriages and hurried pedestrians, it was possible to spot, just barely, a triangular gable crowning a tall central window that looked onto the *cour d'honneur*, the initial courtyard that welcomed elite visitors. Classical in both form and allegory, the gable's bas-relief depicted a cherub holding in his hands a mirror, a sphere, and a compass—the tools of early science. An owl, the sacred companion of Athena, goddess of knowledge, sat proudly at the child's feet. Only the most privileged and learned elite, however, would have the opportunity to admire the geometrical sundial that had been painstakingly carved into the facade of a smaller, more intimate courtyard tucked far away from prying eyes.

Beginning in 1653, the nobleman provided the members of his private "Montmor Academy" with every resource they could imagine: ample space, access to instruments, an extensive library for research, and—of course—full bellies. Montmor's scientific meetings were preceded by private feasts that quickly became the talk of Paris. Long rows of tables dressed with crisp linens lined the perimeter of the second-floor reception hall, the same hall—and likely the same tables—where Denis would many years later perform his infamous blood transfusion experiment on the mentally unstable Mauroy. Each place setting was graced with a perfectly polished silver platter, accompanied by a knife and spoon. (Forks had not yet decorated even the most sophisticated of tables, where fingers often still were the utensil of choice.)[10] The members of Montmor's armylike waitstaff, assigned no more than two guests apiece, hovered obsequiously behind their charges, anticipating and attending to their every need. Dinner tables were crowded with decorative tureens brimming with rich soups, platefuls of roasted pheasant, cheeses both creamy and hard, and decanters overflowing with wine from the Montmor family vineyards. As in most elite Marais households, Montmor's domestic staff relied on regular deliveries of foodstuffs from the nobleman's vast country estates. The nearby markets at Les Halles, with their muddy, smelly, and tight aisles, were where the lower classes shopped. They were not for a family as refined as Montmor's.

In this wealthiest district of the capital, riches from the provinces as well as from much more exotic locales—India, Africa, South America, and New France[11]—were delivered directly to the ground-floor kitchens. In short order these rare delights quickly made their way to the table in a spectacular culinary display. With the appropriate balance of deference and nonchalance, Montmor's smartly dressed household staff presented a selection of rare delicacies from travels far and wide. On the sweeter side of things, chocolate was one of the novelties of the moment.

Chefs in bustling kitchens of imposing private residences compared techniques for drying and roasting precious handfuls of cacao beans harvested in the New World. They debated, with great enthusiasm and pride, the exact amounts of cream, sugar, and vanilla—itself a recent import to Europe—necessary to bring out the fullest flavor of this exotic delight.[12] But the greatest find from South America had to be coffee, which was served with much ceremony in delicate, hand-painted porcelain cups imported from China. And one thing was certain: At the equivalent of almost four thousand dollars a pound, there could not be a better sign of Montmor's wealth and largesse.[13]

In the first years each of the weekly meetings seemed to announce some new scientific discovery. But with the accumulation of highly trained and achingly brilliant minds comes a stunning array of unyielding egos. Fireworks exploded with regularity in the formal meeting rooms, around the dinner table, and in the library and adjacent halls. On more pleasant days shouts bounced off the garden walls—fights that were exacerbated, no doubt, by the copious amounts of wine guzzled by the participants. Their scuffles centered on a single question that textured all scientific endeavors in this moment of upheaval, this moment of scientific "revolution" that was pecking away at established worldviews. It was a question that demanded bellicose philosophers to put all their cards on the table and decide whether they would continue to reside in the comfort of the past or to leap boldly into the future. The question was this: Is truth ultimately knowable?

It was a dilemma that split late-seventeenth-century thinkers into two camps, and each was aligned with a major philosophical figure. Montmor had shown his hand early, when he set his sights on attracting internationally renowned scholars on whose coattails he could ride, and whose presence at his Marais home would draw intellectuals from near and far. He started with René Descartes. The philosopher's arguments were bold and unconventional. For Descartes an understanding of the mysteries of the natural world, as well as of God himself, was attainable through a rigorous method of careful reflection and experimental observation. Descartes outlined a four-step process for reasoned inquiry that laid the foundation for evidence-based modern scientific practice—a road map that would allow the individual to shake the mind free from all doubt.

Montmor had invited Descartes to his home in the Marais and, as he was so practiced at doing, wined and dined the philosopher as if he were royalty. The nobleman had every hope that his offer to grant the discriminating Descartes full use of his country home would help seal the relationship; this would make the Montmor residence, in the city or on the bucolic outskirts of the Parisian countryside, a required stop for any intellectual worth the name. To his bitter disappointment, and for unknown reasons, Descartes declined.[14]

Though Montmor was attached to the ideas of Descartes, and though he would remain so over the course of his life, he did not intend to let the man's refusal kill his dream of forming an academy for natural philosophers. He turned his sights next on the aging Pierre Gassendi, Descartes' philosophical enemy. For Gassendi truth could never be anything more than contingent, uncertain.[15] He believed that the material world was an amalgamation of invisible and indivisible particles. In contrast to Descartes' plenum, in which particles occupied every bit of space seen and unseen, Gassendi posited instead the presence of gaps and holes.[16] These gaps and holes prevented us from mastering the mysteries of nature. Something had to be present to fully understand it: Certainty cannot follow from emptiness.

As probing a philosopher as Gassendi was, he was a gentle

man who did not know what it was to get angry. Indeed, he preferred walking the gardens with Montmor's children to sparring with colleagues.[17] Gassendi's calming presence had done much to ensure respectful interactions among the guests at the private academy. When the philosopher died at the nobleman's home on October 24, 1655, one thing became exceedingly clear to Montmor. He would need to find another intellectual star—and quickly. As he had done with Descartes and later with Gassendi, Montmor set his sights on luring a new luminary to his private academy.

But Montmor would have to look well beyond the borders of his own country this time. Marin Mersenne, René Descartes, Blaise Pascal, and Pierre de Fermat—Gassendi was but one of several French giants of the scientific revolution who had died in just a matter of a few years.[18] Instead the Dutchman Christian Huygens would soon take a place among the most brilliant mathematicians and astronomers of the scientific revolution. And Montmor would not be shy in taking credit for his role in establishing Huygens's reputation in France.

The young Huygens had first been introduced to the Parisian scientific community in the halls of Montmor's home and quickly became a regular at the academy's weekly meetings. Montmor's efforts to gain Huygens's trust and friendship—no doubt through the promise of social connections and monetary resources—had paid off handsomely. And thanks to Montmor's generous support, Huygens would soon solve a riddle that had dogged even Galileo.

As Galileo had noted with surprise nearly four decades earlier, Saturn underwent an odd metamorphosis from a three-bodied form to a single, elliptical form. The Italian astronomer pondered in disbelief Saturn's baffling change in appearance: "Now what is to be said about this strange metamorphosis? Perhaps the two smaller stars have been consumed in the manner of sunspots? Perhaps they have vanished and fled suddenly? Perhaps Saturn has devoured his own children?"[19] Saturn continued to haunt Galileo for over thirty years. In the summer of 1616 he wrote a letter to the prince of Tuscany. It contained an important correction to his original three-body assertion: Saturn's two "companions" were not small, perfectly round globes. They were much larger bodies in the form of half "eclipses" or *anses* (ears) that sat alongside the perfectly round shape of Saturn.[20] Instead of rings Galileo had seen ears, and this description held fast for the nearly forty years that followed.[21]

Huygens speculated instead that something was fluttering around Saturn. "The ears of Saturn," he wrote in 1658 in an encrypted letter intended for the Montmor Academy, "can be nothing other than what I put forth in my anagram":

a c d e g h i l m n o p q r s t u

6 5 1 5 1 1 7 4 2 9 4 2 1 2 1 5 5

The numbers indicated how many times each letter appeared in the enigma, which when rearranged formed the words *Annulo cingitur tenui, plano, nusquam cohaerente ad eclipticam inclinato*: "Saturn is encircled by a thin, flat ring, nowhere touching, inclined to the ecliptic."[22]

Montmor sent Huygens an effusive personal note expressing his gratitude for the astronomer's decision to share the news in the academy first, and indicating his great hope that he would continue to see the assembly as the best place to break the news of other discoveries.[23] Seemingly overnight Huygens had transformed from a starry-eyed young university student to the darling of the Paris scientific community. Where he aligned himself, success was sure to follow.

Montmor was deeply pleased with his own success in recruiting Huygens to his academy. The nobleman often admired a telescope that sat proudly on a gilded stand in front of the room's central window. Galileo's own,[24] the telescope was just over four feet long and was covered in red morocco leather. It had been in Montmor's academy that Huygens had topped Galileo. And now Montmor was more certain than ever that the scientific glory of France—and his own reputation as ultimate patron of knowledge—would continue to be made within the walls of his palatial home.

Chapter 7

"HOW HIGH WILL HE NOT CLIMB?"

In her late-seventeenth-century novel *La Princesse de Clèves,* Madame de La Fayette offers a stark portrait of court life in France. "The Court gravitated," she wrote, "around ambition. Nobody was tranquil or indifferent—everybody was busily trying to better their position by pleasing, helping, or hindering someone else."[1] Social and financial success had long been contingent on variables such as rank, title, marriage, physical attractiveness, and personal ties—both visible and secret. Yet it was now the handsome young king, Louis XIV himself, who served as the ultimate arbiter. He could, at a whim, swiftly alter one's place in the hierarchy—for better or, often, for much worse.

A young man with keen eyes and a strong, square jaw, Louis XIV bore the angular nose that marked him unquestionably as a member of the illustrious Bourbon royal family. His every movement exuded an easy confidence, proof that he trusted fully in the legends that encircled him since birth. He had been named Louis the God Given (*Louis-Dieudonné*) in recognition of his miraculous birth—which took place following two decades of infertility for his parents, Anne of Austria and Louis XIII. It was a birth that had assured the continuation of a fragile monarchy, and Louis XIV joined the populace in believing wholly that he was indeed God given.

As a boy the Sun King had actually shown very few signs of regal behavior. He was socially awkward, tongue-tied, and bashful. Hushed whispers ran through the court that he was "dim-witted" and unfit to lead.[2] But the years that followed were turbulent and forced the young king to grow up quickly. A civil war threatened the authority of the monarchy and had pitted his mother, Anne of Austria, now queen regent, and Prime Minister Mazarin against the troublesome nobles. In the dark of night in 1648, the golden-curled and doe-eyed Louis was shaken awake from his bed in the Louvre by frenzied members of the royal family. They told him little, but their hushed and panicked voices made it clear to the ten-year-old that something was very wrong. Louis, his mother, and the nurse holding Louis' young brother were shuttled into a carriage. He heard frightening words like "plot," "kidnap," and "murder." It was much more than he could take in fully, but the net effect was sheer terror.

The carriage and its precious royal cargo made it safely to the city's outskirts in Saint-Germain-en-Laye, where Louis and his displaced entourage slept uncomfortably on beds of hay and worried about how long they could make the food rations last.[3] There was talk of riots in the city and barricades that stretched as far as the eye could see. For nine months Louis listened as countless tales of unrest in Paris were exchanged in the darkened rooms of their uncomfortable exile. A Paris judge christened the uprising the Fronde (the French term means "slingshot"). Like David the nobles had dared to rise up against the Goliaths of the monarchy. And it looked at times as if the giant—represented by the body of a young child—would be felled.[4]

Louis learned quickly that he was central to French life and, for this reason, extraordinarily vulnerable. By 1653 Paris had grown weary of nearly a half decade of fighting, and yet another war with Spain loomed. France reunited once again around its king. The new king had no patience for nobles who did not understand his divine right to power. But the young monarch vowed never to allow the graying, privileged nobles another chance to bring the crown to its knees. He would have them on theirs first. And those men who did not understand—or refused to recognize—these sea changes on the horizon would soon be swept up into the waves and left to drown.

Montmor would have done well to heed the signals; the fate of Nicolas Fouquet should have been warning enough. For more than a half century, France had been ruled through a de facto partnership between the monarch and his ultrapowerful prime minister. So when Prime Minister Mazarin died in March 1661 after months of suffering, one question and one question alone preoccupied Parisian nobles: Who would succeed Mazarin? Near the top of the list was Nicolas Fouquet. Fouquet had been the superintendent of finances for nearly eight years. His unquestioned loyalty to the throne during the tumultuous years of the Fronde had earned him the coveted treasury post in 1653. In this role he moved money in and out of the royal coffers and negotiated advances and loans in order to secure whatever sums Mazarin and the king needed. When times were tight Fouquet borrowed handsomely on behalf of the state and secured the debts against his own personal property—charging high interest for his services.[5] To be sure, Fouquet came from a respected and reasonably well-off family. But there was no way family fortune alone could explain the magnificent material goods with which he surrounded himself, nor was there a clear financial explanation for the expensive and sumptuous residence that he was building in the countryside just south of Paris.

Vaux-le-Vicomte sat on a property that extended thousands of acres; three entire villages had been torn down and assimilated into the sprawling estate.[6] In his quest to build a home that would surpass anything seen in Europe, Fouquet commanded an army of more than eighteen thousand craftsmen over a four-year period. He recruited the very best masons, painters, sculptors, gardeners, tapestry weavers, bricklayers, and hydraulic engineers. Each worked under the careful eyes of the brilliant artists Le Vau, Le Brun, and Le Nôtre. The trio would later gain fame for their work on the king's even-larger palace at Versailles—but for now, these masters belonged to Fouquet.

A literary patron as well as a shrewd financier, Fouquet opened his estate to a handpicked coterie of the period's most famous artists and writers. Strategic and ambitious to a fault, Fouquet understood well the power of art and literature. He invested in writers much as he handled his financial transactions. He lent his support to artists with the clear expectation that their debt would be paid back in full, accompanied by a hefty profit. To show their gratitude authors offered flattery in print that would have made any other man blush—flattery that would be distributed far and wide, sealing the superintendent's reputation. Fouquet paid the long-silent playwright Pierre Corneille to write again. He paid Molière to stage plays in the gardens of his estate and hired the composer Jean-Baptiste Lully to create music for the performances. Fouquet paid the poet Jean de La Fontaine to pen detailed accounts of the elaborate parties that took place at Vaux-le-Vicomte.[7]

There was no doubt that Vaux-le-Vicomte had been built on Fouquet's skill at embezzlement. Certainly Fouquet was not the first, nor the last, state official to skim funds; prime ministers—

including the celebrated Richelieu and Mazarin—and other members of government had lined their own pockets for years. Yet the coat of arms Fouquet put on display throughout the estate celebrated perhaps too openly his ill-gotten gains. In the Anjou dialect of his home region, *fouquet* meant "squirrel." Throughout Vaux painted squirrels sat proudly in the center of laurel wreaths—the symbol for glory and victory. The wreaths themselves were held up by two lions—no doubt a peremptory reference to the support he anticipated from the lionlike Louis. And woven among the squirrels were scrolls carrying Fouquet's personal motto: *Quo non ascendat?* "How high will he not climb?"

For nearly six months Fouquet steadied his focus on preparations for a singular event: a feast to end all feasts, a celebration of his glory alongside the king. On the evening of August 17, 1661, Fouquet led the king and his court on a tour of his majestic estate. Ladies in silk-layered gowns and gentlemen in knee pants and brightly colored topcoats strolled from room to room. Elegance marked every inch of the château, resplendent with hundreds of tapestries, hand-painted wallpaper, gilded ceilings and chandeliers, richly carved furniture, and sculptures that rivaled those in Rome. Fouquet ushered Louis to the even more elaborately decorated king's chambers. By the time the monarch emerged from a brief rest, the skies had darkened and the stars had just begun to make their appearance. The court reunited in the oval-shaped central salon, which Le Brun had designed to be a "palace of the sun." Statues representing the seasons and the constellations circled the upper walls. In the center of the cupola, where Apollo held court, there was also a squirrel.

In the wee hours of the night, after the second dinner of the evening, Louis bade his host good-bye. Given the hours-long journey ahead of him, the king could have chosen to settle into his comfortable rooms at Fouquet's estate. But he had seen enough, his patience had been tested, it was time to leave. Without warning the sky exploded in light and was just as quickly shrouded in smoke. Amid shrieks of fear that Vaux had been hit by comets or cannons, two horses leaped up, dragging the queen's carriage on its side behind them. Trying desperately to escape the fracas, the horses plunged headlong into the watery moat to their deaths. Once the smoke cleared and the safety of the king and queen had been assured, the source of the explosions was uncovered. To bid the king farewell, Fouquet had launched enough fireworks from the cupola of his château to vanquish a small army.[8]

In the days and weeks that followed the party, Fouquet waited nervously for news from the monarch, remaining hopeful that his show of respect and celebration would net the appointment of prime minister that he so desperately wanted. By early September he could wait no longer. He summoned his household staff and had them pack his bags in haste; he was headed to Nantes where the king was now in residence. We cannot know if Fouquet turned around to admire his castle-like home as his carriage pulled away. If he did not, he should have: It was the last time he would see it.

So sure was Fouquet that he would gain the king's favor, he had overlooked the venomous Jean-Baptiste Colbert in his calculations. An aloof man with thinning brown hair that framed his dimpled round face, Colbert had little patience for social games. In fact his coldness had earned him the nickname *le Nord* (the North) from the most influential social brokers.[9] He had long been aligned with Mazarin, who held him in the highest regard. Colbert's raison d'être was his work for the glory of the king. Somber and brusque, Colbert worked upwards of sixteen hours a day at the king's behest. The monarchy lay at the heart of his every passion; it was his identity, his nourishment.

And like Louis XIV, he loathed the very thought of scheming nobles.

Just a day before his death Mazarin had warned Louis in a fading voice about Fouquet, and confirmed his trust in Colbert.[10] Taking heed of Mazarin's recommendations, the king quietly assigned Colbert the task of investigating the minister of finance. And what Louis saw at Vaux-le-Vicomte more than confirmed the results of Colbert's investigations: Fouquet needed to be stopped. But the investigation had a secondary, and very welcome, effect for the hardworking Colbert: It eliminated a rival for the king's attention. On September 5—the king's birthday—Louis convened a council meeting in Nantes with his ministers, including the two vying for the king's favors: Fouquet and Colbert.[11] When the meeting was adjourned, the king asked Fouquet to stay behind. Fouquet's heart leaped with joy; the moment had arrived. He had been chosen, or so he believed. Engaging the minister in small talk, the king then nodded to a man standing in the doorway behind Fouquet. The king's chief of security, the real musketeer D'Artagnan, stepped forward. Hardly the boisterous character painted by Dumas two centuries later, D'Artagnan was polite and respectful. With the decorum reserved for the highest members of court, the musketeer escorted Fouquet out of the building and into the hands of the fifty guards waiting in the courtyard.[12]

After a court trial that lasted three years, Fouquet was sentenced to life in solitary confinement—first at the Bastille, and later at the squalid prison of Pignerol, near Turin (then under French rule). And by the order of the king, Colbert stripped Vaux-le-Vicomte of its riches, recruited its primary architects, and coordinated the construction of an even larger, more sumptuous château: Versailles. Fouquet's coat of arms—the ubiquitous squirrel—was left intact. Colbert would later have his own family symbol, a grass snake, drawn in. With the squirrel now pursued on both sides by snakes, the animal's motto was also changed to *Quo me vertam nescio*: "I do not know who[m] to turn to."[13] Disgraced, Fouquet died alone in his prison cell in 1680, seventeen years later.

While the young king was working to demonstrate his power in the most visible of ways, meetings at the Montmor estate had become increasingly contentious. The astronomer Ismael Boulliau described with his usual candor the philosophical standoff taking placing at the Montmor Academy:

> The Montmorians . . . dispute with vehemence, since they quarrel about the pursuit of truth; sometimes they are eager to rail at each other, and jealously deny a truth, since each one, although professing to inquire and investigate, would like to be the sole author of truth when discovered. And if anyone in the course of his hunting find that truth, the others will not in the end share in the spoils of their own free will and pleasure, because each one considers that his own fame and glory has lost something if he grant even a blade of grass to the victor and acknowledge him as the real discoverer.[14]

The permanent secretary of the Montmor Academy, Samuel de Sorbière, had made no secret of his belief in centralized power structures. A devotee of Thomas Hobbes, Sorbière had written essays in which he argued vehemently that "men live more happily under a despotic government than one that is less absolute"—an argument Hobbes had laid out in his *Leviathan*.[15] During the English civil war (1642–51), Hobbes urged citizens to yield their rights to their leaders in order to ensure peace. Without a totalitarian government, conflict reigns and humans are consigned to lives that are "solitary, poor, nasty, brutish, and short."[16] Better to have an effective leader who occasionally abuses the power with which he is entrusted than to endure such a miserable fate.

These principles were woven into the fabric of Sorbière's very being; he was committed to making sure that Montmor would fully occupy his rightful place as unquestioned head of his own academy—both for the good of the institution and especially as something of a Hobbesian social experiment. The academy was, he believed, an example of French society in miniature: Rules, regulations, and a central authority were required to maintain peace and ensure progress.[17]

Several years earlier Sorbière had drafted academy regulations in nine articles and presented them publicly to the group. Sorbière's bylaws declared that Montmor alone would set the agenda for each meeting. The master of the house would also have sole authority to designate at will two colleagues to report on their opinions, uninterrupted, about a subject of the nobleman's choosing. All commentary would be written ahead of time and logged in by the permanent secretary, Sorbière himself. "Any interruption during the presentation," he declared, "would not be tolerated." Only in this way would the academy cease wasting its time on "vain exercise of the mind on useless subtleties."[18] While Sorbière espoused a more dictatorial approach to academy business, the ever-amiable Montmor's approach was instead one of accommodation. Enamored of the intellect of the men he had recruited to his home, Montmor had proved himself nonetheless incapable of maintaining order in his stunning collection of scientific enfants terribles. Now, in the wake of Fouquet's arrest and imprisonment, the Hobbesian Sorbière had found a sure-footed, decisive leader whom he could respect. For as much as the nobles may have groused quietly about the shocking treatment of Fouquet at the king's hands. they learned quickly the dangers of testing the king's patience and kept their discontent quiet.

During his tenure as permanent secretary, Sorbière had always paid attention to the direction of the political winds. Out of his newfound admiration for the decisive young king in the wake of the Fouquet affair, he shifted alliances. On Tuesday, April 3, 1663, Sorbière stood before the Montmor Academy. He began by thanking Montmor for his years of dedication to scientific inquiry. "It is certain," Sorbière proclaimed, "that our illustrious moderator was the first to excite in Paris the studies that we have cultivated here, the curiosity that we have had for the works of God, and our desire to move human industry forward." Pausing for effect, Sorbière then launched into what sounded very much like a eulogy. "We can only hope that he will continue to show his commitment to his glory and for the public utility by allowing the Academy to pass into the hands of the sovereign."[19]

"The king is young," Sorbière told the academy. "He has a good soul, and he has shown that he is open to the idea of creating a general Academy where we might continue our work." Scientific learning and discovery, Sorbière argued, were pivotal to the king's nationalist program. They were too important to be left to flounder in private academies. Moreover, the idea of allowing noblemen like Montmor to set the agenda when it came to inquiries into the secrets of the natural world would not only be counterintuitive to the Sun King's efforts to consolidate his power—it would stunt the very progress of science. Sorbière made sure that a copy of his speech found its way to the Louvre. Attached to it was a letter to Colbert. It contained, of course, the familiar sycophantic praise for which Sorbière had become infamous: "You will see that my harangue of April 3," he wrote, "could lead to something important for the public if it is considered by those who work for the ornamentation of France." He had no hopes of securing personal gain in sharing the speech, Sorbière claimed disingenuously to Colbert. He was most interested in ensuring that he and other natural philosophers could focus on the "business of the sciences."

Colbert swatted Sorbière away like a pesky fly. He did not trust Sorbière and found his self-interested requests a nuisance. The prime minister refused to recommend the idea to the king on financial grounds as well. A fiscal conservative, Colbert found himself signing off on more pricey expenditures than he could bear. The construction of Versailles was in full swing. At every turn the king expanded the blueprints and demanded that the palace be built with the most renowned artisans and with the most expensive materials possible. The king had already made plans for a seven-day-long spectacle to celebrate his beloved mistress, Mademoiselle de La Vallière, which meant that construction became more frenzied and exorbitantly costly as workers rushed to meet the summer deadline.

While Colbert seemed initially lukewarm, Sorbière's speech rang clearly in the ears of the scientists at Montmor's academy. It was true that they would remain forever grateful to Montmor and his willingness to offer "the infinity of machines and instruments" and, of course, his legendary banquets.[20] But even the well-heeled son of *Montmor le Riche* could not offer all that was required for scientific progress. "In truth, Messieurs," Sorbière said with conviction, "only Kings, and rich Sovereigns, or some wise and wealthy Republics could undertake to outfit a Science Academy engaged in continuous experimentation." Until then "our Mechanics will remain imperfect as they are, our Medicine will be blind, and our Sciences will teach us with certainty only that there is an infinity of things about which we know nothing." Pulling no punches, Sorbière took special aim at the hopes Montmor had pinned on astronomy. "Just think of the space needed for observation of the stars, and of the size of the apparatus necessary for a forty-foot telescope. . . . Was not Tycho Brahe forced to build his Uraniborg, a castle not so much for lodgings as for the making of celestial observations?"

Built during the last decades of the sixteenth century, Uraniborg held the title of Europe's first dedicated observatory. It sat on a tiny island close to Copenhagen and had been a gift from the Danish king Frederick II to support Brahe's research. The astronomer was unyielding in his attention to detail and in his insistence that observations not remain in the realm of the general but of the meticulously specific. He insisted that the instruments he designed be checked regularly for accuracy and that descriptions of planetary orbits be made at every point in the orbit. As a result Brahe's detailed astronomical data had been indispensable to Johannes Kepler in his work on planetary motion and his discovery that planets trace elliptical paths. By invoking Brahe's name, Sorbière was no doubt trying to capture the attention of the star of the Montmor Academy: Christian Huygens. The arguments must have had traction for Huygens because by early 1663 he also took up the argument for a national academy of sciences. "There is a great desire," Huygens complained to a colleague, "to make some more solid and regular establishment for an Academy than it has had up to this time, and for some time various consultations to this end have been held; with all that however, we make little progress, so that even the most zealous begin to despair of success."[21]

Colbert's initially chilly reception of such petitions shifted dramatically in late 1664 and early 1665, when the stunning comets many claimed were responsible for London's miseries hit the skies and rattled all of Europe. In Paris astronomers and lay observers alike stayed up into the wee hours to marvel at the stunning light in the sky. Comets brought change, of this there was no doubt. Even the musketeer D'Artagnan was said to have spent an entire night awake, staring at the comet and perhaps wondering what its arrival meant for the work of his squadrons.[22] The astronomers pressed their case and argued that a large observatory needed to be built so that France could master celestial secrets instead of gazing up at the skies in fear.

Colbert was now clearly in the mood to listen. For an insider's perspective about what was happening in the scientific community, Colbert turned to Jean Chapelain, a longtime friend of Montmor and a dedicated member of his academy. The graying Chapelain knew well enough that friendship mattered little when one was called to the king's service. By July 1665 Chapelain had given Colbert a list of eighty-two exceptional men worthy of royal protection: Montmor was not one of them.[23] Yet at the top of Chapelain's list was Christian Huygens, the star of the nobleman's private academy.[24]

If comets had filled the skies in 1664 and 1665, 1666 was marked by a stunning solar eclipse on July 2. Now fully convinced of the importance of astronomical research, Colbert offered his own home for observations. Huygens, Adrien Auzout, and Pierre de Roberval—all former members of the Montmor Academy—joined several others at Colbert's residence at the crack of dawn. They arrived bearing two telescopes, a sextant, a pendulum clock (which had been invented by Huygens), as well as other instruments.[25] It is not certain whether Colbert himself joined the men, but it had become exceedingly clear that he had joined their cause—and if he wanted full use of his home again, a different setting needed to be found.

On December 22, 1667, the first official meeting of the French Academy of Sciences took place without fanfare. The academy had received the official approval of Louis XIV—and early science was now an endeavor of the Crown, rather than something bankrolled by wealthy private patrons. The academy was given full access to the king's personal library at 8 rue Vivienne, not far from where the nineteenth-century Opéra Garnier now stands, and just a few doors from Colbert's own home at 2 rue Vivienne.

Prime Minister Colbert had made good on his promises to provide royal support for science. The private academies, such as Montmor's, had outlived their usefulness, and the academicians needed resources that only a king could provide. In exchange each member knew that he now served at the pleasure of Louis XIV—who lavished handsome rewards for unquestioned loyalty, and equally elaborate punishments for anything less. Huygens was offered a financial arrangement that he could not refuse. He enjoyed an annual stipend that was nearly four times what the average French academician earned—plus spacious living quarters in the king's library.[26] The monarch also made a promise that the housing arrangement would be temporary. To accommodate both the research needs and creature comforts of the country's now-premier astronomer, an observatory rivaled by no other would be built at a location of the academy's choosing. Huygens had long stood at the center of the Montmor Academy. Now he was on the king's payroll.

The Montmor Academy "had ended forever," wrote Huygens. "However, it seems that from the wreckage of this one another may be born."[27] Montmor had been betrayed. He had been undermined by Sorbière, surpassed by Huygens, and abandoned by his own academy. There is little trace of Montmor between the troubled exodus that followed Sorbière's speech in 1663 and the official establishment of the Academy of Sciences by Colbert in December 1666. Did Montmor step aside because he understood the futility—and great personal risk—of competing against Louis XIV and his royal resources? If he did, it was not for long.

As Montmor watched the ever-confident Jean-Baptiste Denis transfuse the blood of a young healthy dog into a mangy old one on the banks of the Seine, the nobleman felt confident that Denis would soon become a celebrity in Paris. As he had with Gassendi and Huygens, Montmor resolved to provide the transfusionist with everything he needed to lock in his fame. And thanks to the transfusionist, perhaps the nobleman could compete with the king's Academy of Sciences after all.

Chapter 8

THE KING'S LIBRARY

In the months that followed the demise of Montmor's private academy, one of the principal tasks of the newly established French Academy was to use science to the strategic benefit of the country and the glory of the king. For the academy's geographers, astronomers, and mathematicians, this meant developing surveying techniques to aid the king's armies in their conquests. The job of engineers and physicists was to develop better gunpowders, water pumps, and visionary machines for travel and production. For the physicians it was time to compete fully with the English in the biomedical realm. That meant engaging, for better or for worse, in the blood wars. While triumph in the medical realm would not be as spectacular as a hard-won victory on the battlefield, it would nevertheless be proof of France's superiority over England.

The Academy of Sciences had just begun setting up its laboratory in two wings of the king's library. Overlooking a manicured winter garden, the main experiment room was itself a chaotic work in progress. Half-opened wooden boxes were strewn across the room—and were overflowing with everything an inquisitive seventeenth-century mind might desire. Vials of brightly colored powders, acids and sulfur, alembics, mortars and pestles, magnifying glasses and microscopes, buckets and bowls—and, of course, ropes to restrain the ever-present animals who would face these and other tools of science.

The academy members had their choice of any number of exotic beasts that they could dismantle, layer by decaying layer. Their subjects came directly from Louis XIV's menagerie at Versailles. Built between 1662 and 1664, this early precursor to zoos was home to more than 123 different types of mammals, which were joined by nearly 239 varieties of birds as well as at least 10 types of amphibians, from chameleons to crocodiles.[1] Culled from locales around the globe, the most unusual creatures were more than curiosities; they were jaw-dropping marvels whose odd bodies and deeply colored plumes, fur, or scaled skins tested the limits of the imagination. In this era when tomatoes, coffee, and chocolate had only begun to enter France by way of new travel routes, the very sight of semimythic creatures like a wild lion or an elephant could render a Frenchman speechless.

The animals themselves did not usually fare well. Many were not suited to cold, damp French winters; others starved and died as the result of neglect or were deformed by spending days on end trapped in small cages.[2] Versailles' loss, though, was the Academy of Science's gain. When an animal died, its carcass was delivered immediately to the physicians and anatomists at the king's library or, sometimes, to an academy member's home. As might be imagined, dissection rooms were rarely pleasant spaces, and the Academy of Science's laboratories were no exception. The English traveler Martin Lister wrote later in the seventeenth century that "a private Anatomy room is . . . very irksome if not frightful: Here a Basket of Dissecting Instruments, as Knives, Saws, &c. and there a Form with a Thigh and Leg flayed, and the Muscles parted asunder: On another Form an Arm served after the same manner: Here a Trey full of Bits of Flesh."[3] Shipments of eaux-de-vie, clear and potent spirits usually distilled from fruit, regularly accompanied deliveries of royal animal cadavers. Used to flush rotting body cavities, the spirits also did double duty as a hand cleanser and were often drunk by the anatomists as a way to steady their nerves and stomachs.

Camels, porcupines, lions, monkeys, ostriches, and chameleons: Claude Perrault dissected them all. The steely-eyed Perrault had been tapped by Colbert to explore the truth behind English claims of transfusion. Perrault was less than thrilled at the idea of the newly formed academy becoming involved in blood trials. Like many medical men in Paris, he had been deeply troubled when news of the English transfusion experiments had crossed the Channel into France. Instead of letting blood *out* of veins and arteries as doctors had been doing for centuries, doctors were now being asked to consider ways to put blood *in*. Scandalous and counterintuitive to an extreme, transfusion would be a very hard sell.

As a graduate of the medical school at the University of Paris, Perrault was hardly predisposed to innovation. The most conservative medical school in Europe, the University of Paris had a long-earned reputation for being militantly attached to the theories of Galen and Hippocrates and for battling tenaciously against rival ideas. As every physician who had ever trained at the University of Paris knew without a glimmer of doubt, Galen was more than just correct in his teachings; his philosophy was "unimpeachable," perhaps even "divinely inspired."[4] Bearing the weight of all medical knowledge, Galen's conclusions were like the columns of an ancient monument: To question Galen was to risk demolishing the very temple of medicine.

As Perrault had during his own university studies, Parisian medical students still repeated by heart and without question that the body's inner workings were the result of a "cooking" process that depended on heat. Food could not be transformed into chyle without heat. Without chyle, there could be no blood. And without blood, men would be unable to create semen, and women could not produce breast milk (Galen had argued that both were produced from blood). And if heat produced these key products for reproduction and life, then severe chill meant illness, infertility—or worse—death.

Claude Perrault's own approach to doctoring followed, chapter, verse, and line, the standard humoral ones he had learned as a medical student at the University of Paris. Claude's traditional methods were put on display when his eldest brother, Jean, fell ill during a fateful sightseeing trip to Bordeaux. While Jean was likely struck by a severe case of typhoid fever, both brothers felt certain that the illness had been brought on by sleeping in sheets that had been dried near roses. The roses, which were known for their cooling effect, had created an imbalance in Jean's humors and had affected his body's ability to retain heat. Jean's uncontrollable shivering was thought to be caused by an excessive loss of heat—which was manifested through a fever—brought on in response to an assault of cold humors. This imbalance could only be recalibrated by an aggressive release of the offending humors through purging and bloodletting.

Claude arranged for a barber-surgeon to administer numerous bleedings to his brother's arms and legs. When bleeding seemed no longer to have any effect, they tried to place leeches behind Jean's ears, but blistering there from other treatments with warming salves kept the leeches from doing their work. Bouillons, enemas, and purgings accompanied each bleeding in a desperate attempt to save Jean's life. And to these were added chest rubs with concoctions of ground pearls mixed with extract of hyacinth bulbs to warm Jean's blood, as well as the placement of gutted pigeons on his scalp to create heat to stave off the shivering. Despite Claude's heroic attempts to save his brother (or perhaps because of them), Jean died a few weeks later.[5]

As a founding member of the Academy of Sciences, Perrault took his duties just as seriously and held just as tightly to tradition in his research as he did in his medical practice. Prime Minister Colbert charged Perrault with looking into English claims regarding blood transfusion. Perrault was not pleased. He found the very notion of transfusion too fanciful, and perhaps even too disturbing, to take seriously. But whatever the prime minister—which meant, by extension, the king—asked of him, Perrault did through the blunt force of will.

Perrault agreed to form a committee to look into the matter. Beginning in January 1666, nearly two months before Denis' public transfusion on the banks of the Seine, Perrault and two colleagues launched quietly into a series of transfusion experiments in the king's library.[6] Barking defensively and straining at their leashes, two dogs were tied to the room's central table. The astronomer Auzout ignored their protests as he rooted through boxes in search of items that might be of use in the experiment. Auzout had worked diligently with Huygens to persuade Colbert and the king to establish the new academy, and he was eager to help in any way he could to get the group's research off the ground, even if it meant meddling in blood. Stepping gingerly around the dogs, the surgeon Louis Gayant lifted his large surgical box onto the table and slipped a stiff blood-stained apron over his embroidered clothes; a somber Perrault tossed another log in the fireplace.

Once Auzout confirmed that the dogs were strapped down and their muzzles tied, Perrault and Gayant readied their scalpels—

just as their English colleagues had done. Again high-pitched cries of pain echoed against the cavernous ceilings of the library quarters as the two men sliced into the animals. They quickly inserted the tubes into the vein of one dog and the artery of the other, uniting them exactly as the Royal Society had done. Blood rushed copiously out of each of the dogs, into the cannulas, and onto the table. Perrault's eyes narrowed as they traced the blood dripping to the floor. Staring transfixed in annoyance and confusion, he felt his own blood rush to his face in frustration. Nothing, it seemed, had made its way into the transfusion tubes. The experiment had failed.

Perrault watched in mounting anger as the dogs took their last breaths. He and his colleagues had just failed to pull off an experiment that their competitors, the English virtuosi, claimed was possible. Now he would have to repeat this bloody procedure—and he was not pleased at the thought.

For the University of Paris–trained Perrault and the Academy of Sciences of which he was now a part, the inability to replicate the transfusion experiments meant clearly that the English had exaggerated their claims, or even perhaps that they had lied. Moreover, the failed trials confirmed that the traditional teachings that had glued the French medical community together for centuries would remain intact. In fact this outcome may actually have been determined—or at least wished for—long before the trials at the Academy of Sciences began. Perhaps it was not possible to replicate the English transfusion experiments because, simply put, transfusion itself was not possible.

But still, despite his misgivings about transfusion, Perrault was unwilling to give up. Two days after the first unsuccessful experiment, the three men met again. Growling, Perrault made it clear that he and his colleagues would not simply repeat the English trial. The French Academy would modify the entire experi-

FIGURE 12: Gayant's memo to the Academy of Sciences included hand-drawn sketches of canine-to-canine experiments performed in the French Academy of Sciences during the early months of 1667.

ment and outdo their competitors once and for all. They would do more, and better. The Royal Society had focused on creating a one-way transfusion, from donor to recipient. With his colleagues Perrault rigged a system whereby blood would flow both ways. Each of the metal tubes inserted into the dogs contained

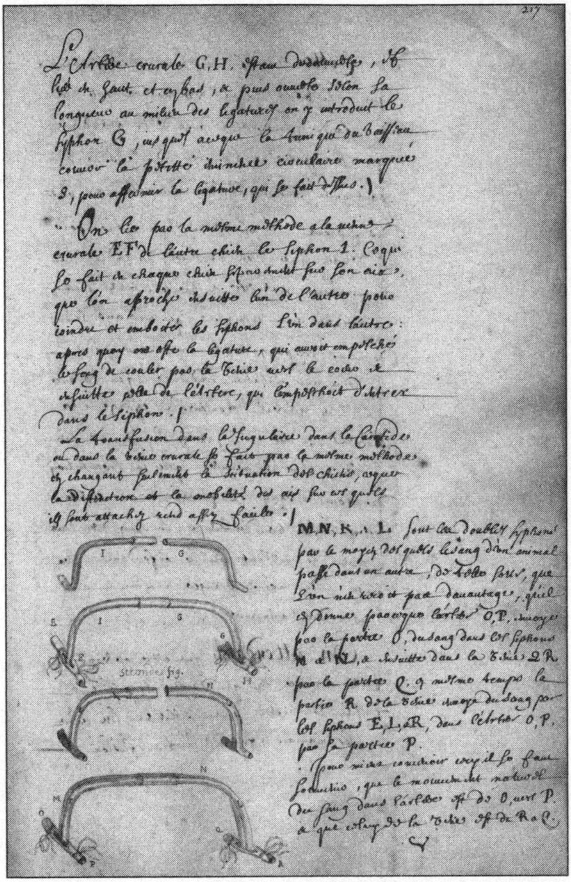

FIGURE 13: Perrault's system of custom-made double-flow transfusion tubes (January 22, 1667).

an uptake cannula as well as an output cannula. That is, each dog would at once be a donor and a recipient. They would, effectively, trade blood. A skilled and trusted metalworker was on-site at the king's library, and he followed the scientists' drawings to the letter as he fabricated special tubes.

This time the transfusion seemed to work. The group marveled as they watched one dog's vein beat rhythmically with the artery of the other dog. It was short lived, however; one dog died almost immediately later. Upon dissection of the recipient, clumps of clotted blood were found in the right ventricle of the recipient's heart. As for the other dog, it lived; but it remained morose and feeble long after the experiment—hardly the spirited pup the Royal Society had promised it would become.

The men were likely seeing, but did not know it, evidence of an antigen reaction. Unlike humans, who have only four blood types, dogs have more than a dozen possible blood types. The likelihood of a blood incompatibility reaction is unpredictable. When transfused with the wrong type of blood, some dogs will show no clear reaction. For others, it can be fatal. It is possible that Perrault's recipient dog was especially sensitive to a foreign blood type. Yet there is also another possible explanation for the dog's death. We do know that in later experiments, Perrault was reusing his dogs. He also performed experiments that he did not always record. If he performed other experiments before this one and reused his dogs, this would go a long way toward explaining the academy's failures in replicating the English experiments both in this specific experiment and later ones. If the recipient dog were given the wrong type of blood more than once, a severe, or even deadly, reaction would have likely followed.

Perrault, Gayant, and Auzout continued their work between January 22 and March 21, 1667. The dogs were dying less often—but the men remained doubtful. How could they be sure that any blood was actually being transfused? In order to confirm the transfer of blood, their last experiment took place on a scale. Weights were stacked in a pan along with the lighter dog in order to balance it with the heavier dog. Once everything was in place, they began the two-way transfusion. The scale rose on one side and lowered on the other. Then suddenly it rebalanced itself and began a fluid dance of up-and-down movements. All told, one dog had received five and a half ounces of blood; the second received just over six ounces—about two-thirds of a cup each. Both dogs died shortly afterward.

For Perrault his failed experiments put English claims about transfusion to rest. The physician argued repeatedly that his trials had more than amply demonstrated the "impossibility that Nature finds in accommodating herself to an alien blood." Blood prepared in the body of one animal was simply not able to nourish the flesh of another. His own experiments, Perrault pointed out, had shown that even the transfer of blood between a single species was fatal. This was precisely the reason for the presence of the umbilical cord and placenta in mammals: "For although the blood of the mother has a great resemblance to that of the fetus, nevertheless it does not at all pass directly from the vessels of the mother into those of the fetus, because it is in reality a foreign blood, and because in this state it cannot be admitted until it is, as it were, naturalized in the placenta."[7]

Like the Englishman Christopher Wren, for whom medicine and architecture were woven together in his plans for a new London, Perrault was similarly fascinated by the relationship between bodies and buildings. As the first to translate the works of the Roman architect Vitruvius into French, Perrault applied Vitruvian dictates of *firmatas, utilitas, venustas* (strong, useful, beautiful) to all that he did on behalf of the king, especially his anatomical work at the royal Academy of Sciences. For Perrault animal bodies had their own geometries, their own symmetries—their own "architecture"—so much so that the physician often displayed his dissected creatures among the very buildings that celebrated the Sun King's reign.

However, unlike Wren, Perrault refused to link blood—and especially heterogeneous tranfusion—with sound architectural

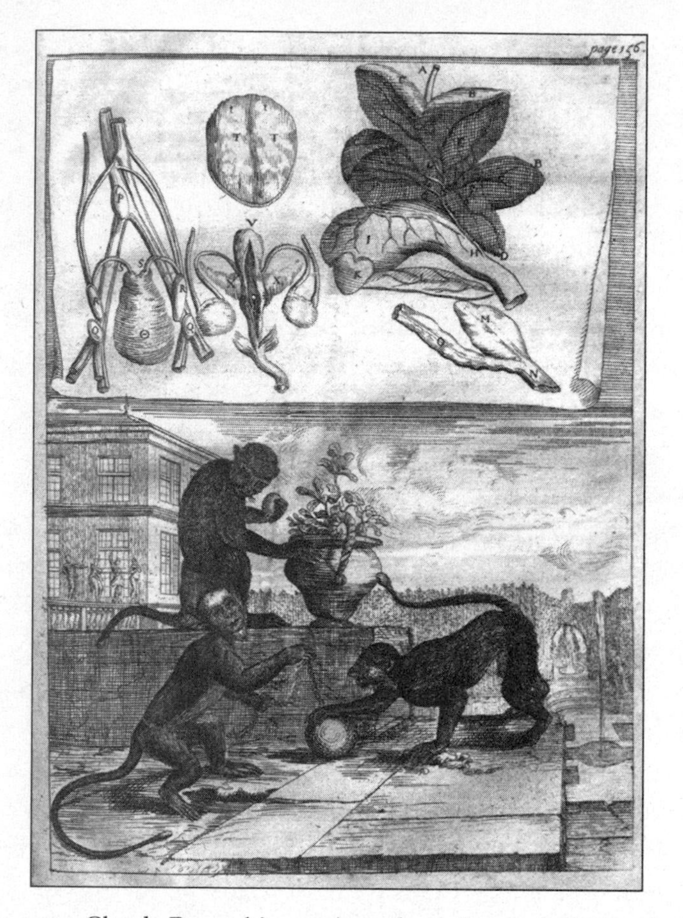

FIGURE 14: Claude Perrault's monkeys, both live and dissected, with Versailles in the background. *Histoire Naturelle* (1676).

design. "Just as the superior construction of a palace," Perrault asserted, "cannot be affected except from materials cut and appropriate to its particular structure . . . so the parts of each animal cannot be nourished except by blood which has been prepared by these very parts." Perrault could not have been clearer about his disdain for transfusion in his architectural references: "The flesh of a dog cannot be nourished and repaired . . . by the blood of another dog, any more than the stone which is cut for an arch can serve either for the construction of a wall or even for another arch than that for which it was cut."[8]

With the unyielding Perrault directing behind the scenes, the Academy of Sciences and the Paris medical faculty—Perrault was a member of both—made their stance against transfusion clear. And when Claude Perrault made up his mind, there was no doubting his determination. That was abundantly clear when the physician-architect designed an imposing triumphal arch in the Saint-Antoine quarter to honor Louis XIV. Perrault created an ingenious system of interlocking unmortared stones—each measuring nearly twelve feet high, four feet wide, and two feet deep—which were then meticulously polished until they were fused seamlessly together. The arch was, like Perrault himself, so stubborn that plans to demolish it decades later were nearly abandoned. In the end the entire structure had to be carted away intact.[9]

Chapter 9

THE PHILOSOPHER'S STONE

In his closing remarks on blood transfusion, Claude Perrault admonished the English by explaining that "the method employed by Medea to rejuvenate her father-in-law was less fabulous and more probable"[1] than the claims that had been made about transfusion. And by evoking Medea, Perrault neatly classified the new science of transfusion among the superstitions of past eras. Transfusion had long been associated with antiquity and, more precisely, Medea, the sensual, ravishing sorceress of Greek mythology.

Medea held seemingly limitless powers to seduce, to plot—and to kill. In one famous mythological scene, Medea commanded servants of King Pelias, her enemy, to bring her an old sheep. Knife in hand, she bled the limp beast nearly dry and cast it into a bubbling cauldron. Soon, a transformed young lamb leaped out of the cauldron. Impatiently she turned next to the king's daughters, who hovered in a drugged trance induced by the witch's herbs. On Medea's command, they pounced like ravenous beasts. Mimicking the sorceress's brutal and precise cuts, the daughters deftly sliced open their father's veins and drained them dry. Medea fled the scene, smug with success at murdering Pelias secondhand.

For Medea transfusion was not just a form of deception. When the sorceress chose to allow it to work, it worked—magically and miraculously. Using claims of transfusion to kill her husband's rival, she also used the real thing on his father, Aeson. She stirred together a brew of "a thousand things," wrote Ovid, "and when Medea saw her brew was ripe,"

> She flashed a knife and cut the old man's throat.
> Draining old veins she poured hot liquor down
> Some steaming through his throat, some through his lips
> 'Til his hair grew black and straight, all grayness gone.
> His chest and shoulders swelled with youthful vigor
> His wrinkles fell away, his loins grew stout . . .
> And Aeson, dazed, remembered this new self
> Was what he had been forty years ago.[2]

Perrault may have scoffed at such tales of Medea's transformative magic, but stories of blood transfusion's metamorphoses were alive and well in the early European imagination—and they drove an even-greater wedge into the already tense relationship between doctors and natural philosophers in Protestant England and Catholic France. The move to science from superstition was far from linear in what some now call, perhaps too reverently, the Scientific Revolution. In late-seventeenth-century Protestant England, science and alchemy were not distinct. The elusive quest for the philosopher's stone—the mystical chemical secret that could be used to transmute base metals into gold—was considered the work of "chymists" and alchemists alike. In fact it was not unusual for books with titles like Michael Sendivogius's *New*

Light on Chemistry (*Novum lumen chymicum*) and other collections such as the *The Theaters of Chemistry* (*Theatra chemica*) to contain chapters on transmutation—the calling card of alchemy. Yet if "chymistry" books often evoked transmutation, many "alchemy" books made, interestingly, no mention of the production of gold, as was the case with Andreas Libavius's *Alchemia*.[3]

The "Father of Modern Chemistry," Robert Boyle himself, claimed to have made progress in his efforts to produce "philosophical mercury," a critical component for the creation of the philosopher's stone. At the age of forty Boyle still had a rosy, youthful face. He was a handsome man who, unmarried, preferred to spend most of his time in his laboratories at Oxford or in the experiment rooms he had set up in his sister's London home. There half-finished manuscripts and papers were strewn through his laboratory,[4] competing for space with balances, jars and vials, mortars and pestles, "weather glasses" (barometers) and "thermoscopes." Bottles and handblown glass tubes cluttered the rows of shelves and tables that sat near the fireplace along with several alembics, used to distill the herbs and plants growing just outside in a small "physick garden."

In his laboratory Boyle experimented with—among other things—transmutation. Breaking the code of silence that typically surrounded transmutation experiments, Boyle described how he purified a small amount of mercury by warming it and mixing it with gold. Writing in the *Philosophical Transactions* under the easily decipherable initials B. R., he explained how he placed a small amount of each in his palm and stirred the mixture delicately with his fingers. The concoction became "considerably hot" and within just one minute the gold dust had melted. Henry Oldenburg was among those who observed Boyle in his laboratory and, according to Boyle, confirmed "with his own hands" that the experiment was successful. Boyle's curious treatise on philosophical mercury ended with an exhortation to readers not to push him for further details. He explained that he would "by all means avoid, for divers reasons . . . divers Queries and perhaps requests (relating to this Mercury)." The scientist later noted that his primary reason for not revealing the full secrets behind his work was that he was fearful of the "political inconveniences that might ensue if it should . . . fall into ill hands."[5]

Still, Boyle wondered publicly whether transfusion might be another form of alchemical transmutation—a type of physiological philosopher's stone. In fact the transformative potential of blood transfusion could not have been more intellectually invigorating for the chemist Boyle. In a letter to Richard Lower that was later read before the Royal Society, Boyle drew up a list of sixteen queries, hoping that they would "excite and assist others in a matter . . . to be well prosecuted."[6] In it he acknowledged that the task of solving blood's greatest intellectual puzzles was just too monumental for one man to undertake without help and issued a plea in the *Philosophical Transactions* to other "learned persons" for assistance.

Of Boyle's sixteen questions, six pondered the impact of the procedure on the appetite of the recipient, an idea based on Lower's early theory that blood transfusion could provide a means for intravenous feeding. The remaining questions focused on possible experiments that would probe the full range of potential transmutations. Most notably, Boyle was interested in learning the extent to which blood transfer between beasts of different species would lead, or not, to an eventual change in an animal's behavior and appearance:

> Whether by this way of transfusing blood, the disposition of individual animals of the same kind, may not be much altered? (As whether a *fierce* Dog, by being often quite new

stocked with blood of a *cowardly* Dog, may not become more tame; & vice versa?)

Whether acquired habits will be destroy'd or impair'd by this Experiment? (As whether a Dog, taught to fetch and carry, or to dive after Ducks, or to set, will after frequent and full recruits of blood of Dogs unfit for those Exercises, be as good at them, as before?)

What will be the [outcome] of frequently stocking (which is feasible enough) an *old* and feeble Dog with the blood of *young* ones as to liveliness, dullness, drowsiness, squeamishness, & vice versa?

Whether the *Colour* of the Hair or Feathers of the *Recipient* animal, by frequent repeating of this Operation, will be changed into that of the *Emittent*?[7]

Boyle's proposals surrounding blood transfusion's transformative potential reflected a cultural fascination for hybrid beasts that had endured since antiquity. As early as the first century BC, Pliny the Elder compiled a lengthy encyclopedic account of more than forty different groups of odd and strangely spellbinding peoples. Among the lengthy list of "Plinian Races" were cannibals, troglodytes, and pygmies, as well as misshapen headless men with eyes on their chests. The Middle Ages inherited Pliny's catalog of races, embellishing them with new descriptions of mysterious peoples that were said to roam the earth. One of the best-known accounts came from Odoric of Pordenone, a fourteenth-century Italian traveler. Friar Odoric was dispatched to the East by the Franciscan Order on a mission to convert heathens to Catholicism. Curious and gifted with words, he documented his travels in the form of an adventure tale filled with amazing creatures. In the Nicobar Islands, he claimed to have met a well-dressed and highly organized race of dog-faced people, the "cynocephali." Odoric described his appreciation for their king, who "attends to justice and maintains it, and throughout his realm all men may fare safely."

On an even stranger occasion during his travels, Odoric stopped to rest at a monastery nearer to China. After dinner, one of the monks cleaned the tables of scraps and invited the traveler to feed the animals who lived in the countryside. The two men strolled in the lush foothills. When they reached their destination the monk struck a gong loudly; at the sound the trees rustled, and "a multitude of animals" emerged from a nearby grotto. Odoric marveled at the "apes, monkeys, and many other animals having faces like men, to the number of some three thousand" that surrounded him. Laughing heartily, he asked his companion what these beasts were. The monk replied, "These animals be the

FIGURE 15: Odoric of Pordenone's human-faced animals.

FIGURE 16: Odoric of Pordenone's society of dog-faced men, the "cynocephali."

souls of gentlemen, which we feed in this fashion for the love of God." Odoric quickly disagreed. "No souls be these, but brute beasts of sundry kinds."[8]

Odoric's medieval claims for the existence of hybrid species joined other similarly detailed—and fanciful—accounts by such travelers as Marco Polo, Sir John Mandeville, and many others. And if we are to believe early modern observers, equally strange "monsters" were just as likely to walk among Europeans as they were to haunt the high seas. In 1657 Royal Society fellow John Evelyn described a lovely harpsichord-playing maiden who was being exhibited, for a charge, throughout the city. Barbara Urselin was covered from head to toe with silky soft blond hair and, wrote Evelyn, "a most prolix beard . . . exactly like an Iceland Dog."[9] A hog-faced and cloven-footed gentlewoman named "Mistris Tannakin Skinker" had likewise made an appearance in London, ostensibly in search of a husband, just a few decades ear-

FIGURE 17: *Barbara Urselin (née Vanbeck), a Very Hairy Woman* by Robert Gaywood (1656).

lier. The pig-maiden had, it seemed, no luck in securing a willing groom and moved to France because "although she has a golden purse, she [was] not fit to be a nurse in England."[10]

There has been some speculation more recently that Barbara Urselin and Tannakin Skinker both showed signs of atavism. That is, their bodies exhibited the reemergence of a lost physical trait or behavior that was once typical in remote animal ancestors. A few examples of atavism in humans are the presence of extra nipples, a tail-like growth, or excessive hairiness. However, most medical historians concur that Barbara was most likely suffering from acute congenital hypertrichosis, an extraordinarily rare genetic disorder. And in the case of Tannakin Skinker, her piglike features were more likely the result of a severe facial malformation that allowed the girl to speak only in squeaking noises, giving rise to rumors that became more and more fanciful over time.[11]

In the absence of genetic explanations for such misshapen women, however, early European thinkers had little recourse but to assign either divine or diabolical causes for the existence of "monsters." The Renaissance writer Ambroise Paré classified monsters into categories based on their origins. Some had been willed by God as a divine sign or portent that, if read correctly, could allow the prognosticator a privileged glimpse into heavenly mysteries. Others were the direct work of the devil, human frailty, or both. For example, writers suspected that Tannakin Skinker had been bewitched in utero after her mother had refused to give money to an old beggar woman. And for both Tannakin and Barbara Urselin, speculation swirled that their mothers had themselves been lusty witches who had copulated with the devil (hence Tannakin's cloven hooves) or had, at the very least, been fathered by animals.

Aristotle's notion of the Great Chain of Being had long held that nature was organized according to strict hierarchies, with God and angels at the top, followed by humans, animals, and then plants. But what rung of the *scala naturae* (nature's ladder) did such odd beasts inhabit?[12] Indeed, from the Middle Ages into the early seventeenth century, these "monsters" actually reinforced the natural order more than they subverted it. They served as a most welcome sign that there was indeed a teleological order in nature. As a term derived from the Latin *monstrare* (to point out), "monsters" reminded and reassured humankind that there was, after all, a purposeful organization of the universe that was specifically willed by God. Misshapen humans—whether actually seen or simply the product of an active early European imagination—offered a welcome opportunity to gain access to God's mysteries. If there were odd, imperfect hybrid creatures roaming the world, their purpose was to reinforce the idea that perfection did exist. In these specific instances nature's rules were broken. But for rules to be broken, there first had to be rules. These anomalies were meant to be isolated, examined, and understood not just as an intellectual exercise but as an act of faith that God had a plan for everything—even for those things that made, on the face of it, little sense.

But how could there be both monsters and a divine, inviolable design? This was the most perplexing question of all. A precious few monsters were, as men like Paré could only reluctantly conclude, simply the result of nature gone awry. In the absence of any plausible divine or diabolical origin, stunted or missing limbs could be seen only as the result of insufficient amounts of "seed" during the sex act. Extra limbs or even conjoined twins followed from too much. For early philosophers the only way to make sense of this was to remind themselves that nothing in the universe was perfect other than God. Even man, created in God's image, was only a poor imitation. Everything in the natural world, humans included, was riddled with imperfection—some were just more horrifically imperfect than others. And now it looked as if natural philosophers would soon be able to engineer their own unclassifiable "monsters."

Boyle was not the only one who was fascinated by the possibility of transmutation and the creation of hybrid species through blood exchange. The protestant Queen Christina of Sweden mused that "the invention of injecting blood is all very fine, but I should not like to try it myself, for fear that I might turn into a sheep. But if I were to experience a metamorphosis, I should prefer to become a female lion, so that no one could devour me."[13] Such questions also delighted Englishmen like Samuel Pepys, who mused over pints of ale what could happen if the "blood of a Quaker [were] let into an Archbishop?"[14]

Pepys's playful reference to religion evokes the reformationist mindset that extended into nearly all aspects of Protestant intellectual life, including and especially medicine. "If Luther could break from Rome," suggests one prominent historian, "how could it be impious to demand the reformation of medicine?"[15] And transfusion—associated as it was with the radical theory of circulation and the iconoclastic practices of alchemy—sat squarely in such efforts. For English physicians and natural philosophers, Luther indeed had an equivalent in medical practice and theory: the alchemist Paracelsus (1493–1541), who insisted that everything in the universe—including humans—was, at the core, made up of minerals and metals such as arsenic, lead, copper, iron, and gold. But three in particular dominated them all: salt, mercury, and sulfur. Each part of the human body was controlled by an *archeus* (master spirit) that mixed, stirred, and transmutated these metals and minerals together to promote good health. Using the language of alchemical tools and materials—stills, casks, filters—to describe the inner workings of the body, Paracelsus argued that disease was what happened when something went awry in the "alchemist's kitchen."[16] If disease was brought on by alchemical processes, then the physician must be first and foremost an alchemist to cure it.

Welcomed in Protestant England, Paracelsus was loathed by many in Catholic France—and blood transfusion and talk of transmutation did not help the matter any.[17] With its perceived links to alchemy, blood transfusion opened up old wounds from an ideological battle the Paris Faculty of Medicine had been waging since the late sixteenth century and had every intention of winning now. Paracelsus's irreverent theories had long ago turned the northern and southern faculties of medicine—in particular, Perrault's Paris and Denis' Montpellier—against each other. Hostile to the theories of Paracelsus and his followers, the Paris faculty decreed in 1566 that the use of chemically based remedies was to be expressly prohibited. More open to Reformationist approaches to medicine, the Montpellier faculty promoted regular use of antimony—a Paracelsian chemical cure par excellence—to treat a broad range of illnesses, from plague to paralysis, from asthma to allergies. Antimony, a metalloid, is now used in the production of electronics, flameproof coatings, and enamels. But in the early modern era, it was prized by some for its powerful emetic properties, which could be used to purge the body of other illness-causing minerals and metals. The Italian physician and botanist Pietro Andrea Mattioli had made a convincing argument, at least for Montpellerian practitioners, that antimony's action on the body was similar to its presumed action in alchemical experiments. With the help of antimony, impurities could be freed from gold. Since gold was the most perfect of metals, and man the most perfect of beings on earth, it made sense that antimony could also be used to remove impurities from the human body.[18]

In 1667, the same year as Denis' experiments, parliament reversed the 1566 decree that forbade the use of chemical remedies. The groundwork for the reversal had been laid some nine years earlier, in 1658, when Louis XIV took ill during a military campaign in Flanders. His personal physicians tried every remedy they knew to cure their patient, but it was only after a local doctor administered antimony that the king returned to good health. Following the parliamentary decree, a prominent member of the Paris medical school declared, "These doctors say that a poison is not a poison in the hands of a good physician. They speak against their own experience because most of them have killed their wives, their children, and their friends."[19]

With the advent of transfusion, coupled with setbacks in regard to antimony, the philosophical conflicts between the two faculties reached fever pitch. As Denis forged ahead with his transfusion experiments, he refused to yield to Perrault's declarations—and by extension those of both the Academy of Sciences and the Paris Faculty of Medicine—that blood transfusion was not an acceptable avenue of medical inquiry. As such, the Montpellier-trained Denis represented something of a return of the repressed—a reincarnation of a battle decades earlier that had clearly not been resolved. And now the battle had taken a turn for the worse. Following the suggestions of Robert Boyle and the Royal Society, Denis was preparing to transfuse a variety of different animals one to the other—and soon would be turning to interspecies trials with humans. For men like Perrault and others, the idea was not only unacceptable—it was terrifying. As Perrault himself had warned unequivocally, to meddle with blood could mean only one thing: great peril.[20]

Chapter 10

THE BLOOD OF A BEAST

As the winter of 1667 slowly gave way to spring, Montmor's Italianate gardens began to display a welcome show of color. Clusters of flowers were tucked inside low evergreen hedges that formed compact designs. Along the back walls of Montmor's private haven, the chestnut trees took bud and began to obscure from view the estate's kitchen gardens, which lay hidden behind its main walls. A stableman shoveled mounds of horse manure onto the leek and potato beds while members of the domestic staff plucked insects from the cabbage patch and folded lettuce greens into their aprons. Costumed in the fresh-pressed clothes of the nobleman he aspired to be, Denis walked along a row of cages lining the perimeter of the formal garden. There he inspected the dogs that Montmor's servants imported daily to serve as his experimental subjects.

From the window of the first-floor reception hall, Montmor likely stole a peek at his protégé's work with excitement and satisfaction. Once the gardens had crackled with intellectual energy. Birds had flown out of the trees in panic as the engineer

Pierre Petit had tested the trajectory of bullets shot from guns loaded with different mixtures of saltpeter, and no metal object had been safe from the physicist Jacques Rohault's experiments with magnets. And on cool days the great physician Jean Pecquet, for whom a structure in the thoracic duct (Pecquet's Cistern) was later named, had performed dissections on a host of animals, not to mention a human corpse or two.[1] So many of these men had packed up and headed to the Academy of Sciences that the gardens were now empty, save for Denis and his assistants. But Montmor was sure that if his bet paid off with Denis, the others—the ones who had abandoned him—would soon be begging to return.

With the help of Emmerez and the lackeys Montmor supplied, Denis patiently worked his way through a logical progression of experimental techniques. He transfused pairs of dogs from artery to vein, vein to vein, neck to neck, leg to leg, in dogs "both weak and strong, great and small." By Denis' account the experiments had been successful. Of the nineteen dogs on which they experimented, not one died.

Never one to hide his light under a bushel, Denis announced his success far and wide. With the support of Montmor, Denis submitted written reports of his work to the *Journal des sçavans* and struck up a correspondence with Henry Oldenburg, the editor of the *Philosophical Transactions,* in the hope that his successes would find their way into the influential English journal. In short order this once-unknown young doctor from Montpellier and his cause célèbre had taken center stage in the debates of the European scientific community. And with each day that passed, the outsider Denis earned himself new enemies.

When Denis had finished his last dog-to-dog transfusion, the transfusionist announced confidently to Montmor that he was working on ideas to "drive the business yet a little further." He would, following in Boyle's footsteps, turn next to the possibilities of transfusing animals from different species.[2] On a crisp day in early April 1667, Denis readied his experiment in Montmor's gardens; this time his host joined him. A shipment of several calves and a selection of dogs for Denis' next experiment, brought in by the stablemen, awaited the transfusionist.

Denis inspected each of the animals and chose his next subjects. A large makeshift table groaned as Montmor's stablemen wrestled with one of the young but still heavy cows. After a few minutes they had restrained it by an elaborate system of ropes secured to thick metal stakes pounded into the ground. The animal now lay paralyzed in fear, looking up at its experimenters with large, bulging eyes.

Then they turned their attention to the small dog that would be joining the calf on the table. If the stablemen thought it would be easier to restrain the dog, they were wrong. The dog nipped at and bit anyone who tried to come close. After some wrangling and many scratches, the men had triumphed over the beast. Scalpels in hand, Denis and Emmerez stepped forward as they had done countless times before. The recalcitrant dog fought back as the first incision was made. But its menacing growls soon gave way to loud, rhythmic, high-pitched yelps that quieted slowly with each pulse of blood that poured from the animal's body. Calf's blood flowed through the now familiar setup of metal tubes and into the dog. Denis and Emmerez bled the calf dry, and the animal spasmed as it took its last breath. The dog's breathing was also shallow and labored; it lived but remained weak. By the end of the experiment, both men stood together in the crimson puddle that encircled their worktable.

Denis repeated the experiment two more times, each time transfusing calf blood into a dog. "The animals into whom the blood has been transmitted," Denis reported proudly in the pages of the

Journal des sçavans, "all of them eat as well as before, and one of these three dogs, from whom the day before so much blood had been drawn, that he could hardly stir any more, having been supplied the next morning with the blood of a Calf, recover'd instantly his strength and showed a surprising vigor."[3] Continuing his experiments, he crowded still more animals onto the transfusion table: Three sheep were transfused with three dogs, a young cow with a dog, and a horse with four goats. As usual Denis made certain that news of his experiment was published in the *Journal des sçavans* just days later. The transfusionist failed to make any mention of the English influence on his work, or of Boyle's lengthy memo about the next directions in interspecies trials.

But the English were, like Denis, also hard at it. As early as January 1667 Oldenburg reported to a colleague in a private letter that the Royal Society was "now busy with the experiment of transfusing the blood of one animal into another, either of the same species or of a different one. Whether that will bring about any change in the creature's nature or not will soon appear, unless perhaps winter's severity by causing the blood to become stagnant and thick hinders it."[4] A few months later the physician Edmund King reported that had not seen any species changes—yet—in his own experiments, but did say he had witnessed some dramatic transformations in the health of the animals who received interspecies blood. In the pages of the *Philosophical Transactions* he described how he had transfused more than forty-five ounces of calf's blood into a sheep. The sheep was "very strong and lusty" afterward and was sent back out to pasture without complications. Writing in the same issue, Thomas Coxe acknowledged that he, too, had transfused "an old Mungrell Curr, all over-run with the Mange" with the blood of a young and healthy spaniel. "The effect of which Experiment was," Coxe reported, "no alteration at all, any way, to be observed in the Sound Dog.

But for the Mangy Dog, he was in about ten days or a fortnight's space perfectly cured."[5]

Thus the English were more attached than ever to the idea that transfusing or otherwise manipulating blood could produce transformative results. In fact some natural philosophers were so duped by their own scientific ambitions and eager hopes that they lost sight of their ability to judge the results of their experiences objectively. For example, as King reported in the *Philosophical Transactions*, he infused a sheep with milk and sugar. He claimed the results were anything but "distasteful"; in fact, the animal was "more than ordinarily sweet, according to the opinion of many that ate of it."[6]

As the English continued to press forward with their experiments, Denis knew he had little time to waste if he wanted to maintain his lead in the transfusion race. His work became still more urgent when he learned that the Italians were also now attempting the experiments—and were making some dramatic claims about the effects. During the Renaissance, Italy had been the center of cutting-edge science, claiming Leonardo, Galileo, Vesalius, and many others as its sons. Now, a century later, the Italians had been eclipsed by the French and, especially, the English. Rumors had been buzzing in Rome and in the prominent medical center of Bologna that blood transfusion would return Italy to its rightful place on the world's scientific stage.[7]

On March 28, 1667—just two weeks after the *Journal des sçavans* published Denis' first report of successful canine tranfusions—the natural philosopher Giovanni Cassini performed experiments on sheep. By that May—again just weeks after Denis' cross-transfusion experiments with dogs, goats, horses, and cows—the Italians began interspecies experiments. In his home in Udine a surgeon named Griffoni brought his own spaniel to the operating table. The dog was thirteen years old, was deaf, and had difficulty walking. Griffoni transfused lamb's blood into the dog's veins and left it to recover for an hour, untied, on the table. Griffoni and his colleagues moved to another room following the experiment—and were delighted when the dog reportedly jumped off the experiment table and bounded into the room where the men were drinking and relaxing. Within weeks the surgeon swore that the dog had been cured of its deafness—or almost. The dog returned at least "sometimes at the voice of his Masters."[8]

Not long after, one Ippolito Magnani fine-tuned the tools used in transfusion, rejecting metal tubes in favor of glass ones in an experiment that mixed the blood of two goats with that of two dogs. He was pleased to note that the blood flowed more freely in glass than in metal.[9] And by the fall of 1667 the transfusionist and respected doctor Paolo Manfredi had gained the support of major courtly benefactors including Marie Mancini, former mistress of Louis XIV and niece of the Cardinal Mazarin, his former prime minister, for whom he may have also demonstrated the experiment in the family's sprawling Palazzo Colonna in Rome.[10]

With the competition becoming more intense with every passing day, Denis remained stubbornly convinced that ultimate victory would belong only to whoever succeeded in performing the first transfusion in humans. Finding neither "reasons nor evidence" sufficient to shake his belief in transfusion, Denis felt that the only remaining issue was to decide on the donor for the experiments in humans. Denis immediately ruled out suggestions that it would be best to use the blood of the same species—that is, to attempt human-to-human transfusions. It was, he believed, "barbarous" to shorten the life of one man to extend the life of another. "Many had conceiv'd," Denis explained, "that if ever the transfusion of blood should come to be practis'd upon men, it ought to be done with blood of the same species. . . . But for my part, I am far from that Opinion, and I am persuaded that it will be much more expedient to make use of the blood of other Animals."[11]

The advantages of animal blood were self-evident to Denis. Greater quantities of blood could be drawn from animals because they are often larger than humans. And finally and most important: Animals did not drink, swear, or overindulge their passions. Animals are less subject to the "sadness, envy, anger, melancholy, disgust, and generally all the passions that trouble the life of man and corrupt the whole substance of the blood."[12] Even the blood of young children is less preferable to that of animals because infants suckle breast milk, which was understood at the time to be produced through a distillation of the blood in the breasts. And a mother's milk, like her blood, was subject to the same "corruptions." As Denis explained to Montmor, "Animal blood necessarily has fewer impurities than human blood." Their blood was, in a word, untainted.

As counterintuitive as it may seem to us today, it makes perfect sense that Denis would prefer animals over humans as donors in his groundbreaking blood transfusion experiments. Animal flesh and fluids were prescribed for centuries for every ailment imaginable. Printed and manuscript "Physick Books" contained myriad recipes for animal-based ointments, compresses, tinctures, and capsules to be used in home healing. One common treatment for consumption called for a live cock: "Slit him down the back and take out his Intrals, cut him in quarters, and bruise him in a Mortar, with his Head, Legs, Heart, Liver and Gizard; put him into an ordinary Still with a Pottle of sack sherry." Still more elaborate was a cure for kidney stones:

> In the month of May distill Cow-dung, then take two live Hares, and strangle them in their blood, then take the one

> of them, and put it into an earthen vessel of a pot, and cover it well with mortar made of horse dung and hay, and bake it in an oven with household bread and let it still in an oven two or three days, until the hare be baked or dried to powder; then beat it well and keep it for your use. The other Hare you must flea, and then take out the guts only; then distill all the rest, and keep this water; then take at the new and full of the moon, or any other time, three mornings together as much of this powder as will lie on six pence, with two spoonfuls of each water; and it will break any tone in the kidneys.[13]

Each animal was classified according to its perceived helpfulness in healing specific parts of the body. The flesh of foxes was considered helpful for persons who suffered from pulmonary problems, and their livers were a good nutritional supplement for those with "sweet urine" (diabetes). Beaver meat could be used to supplement the diets of those who had stomach problems or, for women, "womb troubles." Deer were something of a cure-all. Both folk remedies and learned medical manuals touted their ability to cure a variety of maladies: plague, smallpox, mumps, rheumatoid arthritis, cataracts, paralysis, and impotence.[14]

If animal flesh and blood had long been part of the standard medical regimen in his day, Denis found it logical to shorten the route that the blood had to take through the human digestive system and to place it directly into the veins. There was now a real possibility of curing disease efficiently and directly through blood from a donor with qualities perfectly suited to the humors of the recipient—even if that donor was not itself human.

As Perrault mobilized the Paris medical establishment against transfusion, Denis continued to press forward with abandon. Thanks to the now-marginalized-but-still-wealthy Montmor, Denis was flush with funds and supplies to take his blood studies as far as he could. His intentions were clear: He would without delay be the first to try this radical new procedure on humans—and he would use animal blood. And what better choice than a sheep for this monumental experiment? Lamb of God, blood of Christ: Nothing could be more pure.

Chapter 11

THE TOWER OF LONDON

By mid-June 1667 Denis found a good candidate for his radical procedure when he was called to the home of a boy, barely sixteen years old. The patient had suffered from uncontrollable fevers for two months straight, and barber-surgeons had bled the boy more than twenty times, to no effect. Denis does not tell us how he was able to persuade the patient and his parents to submit to the experiment, although, given what we know of similar cases, we can speculate that some payment may have been involved. At five o'clock in the morning—before the teenager had a chance to stir from bed and heat up his already boiling blood—Denis and his barber-surgeon companion Emmerez tied a tourniquet around the patient's arm. They bled three ounces from him: three ounces of the blackest, most putrefied blood they said they had ever seen. On cue, the family butcher brought in a lamb and set to work opening its carotid artery. The patient and the animal were soon linked by rudimentary metal tubes. Denis reported that the young man shuddered as he felt a strong sensation of heat in his arm, a sign of a mild hemolytic transfusion reac-

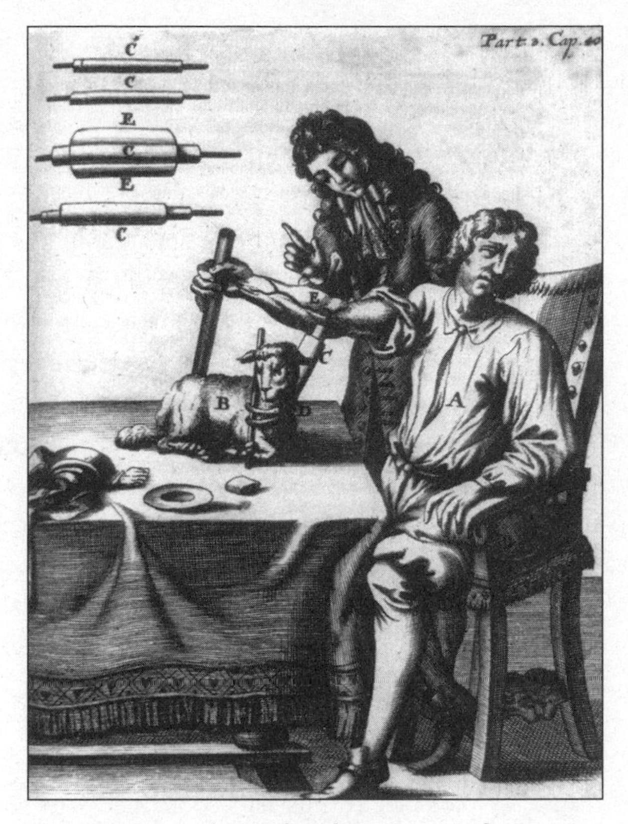

FIGURE 18: Animal-to-human transfusion.
Mathias Gottfried Purmann (1705).

tion. Then, according to Denis, his body relaxed as an immediate feeling of coolness and peace overcame him.

By the next morning the teenager was alert, agile, and seemingly cured of his lengthy illness. Emboldened, Denis paid a healthy, middle-aged man to undergo a similar transfusion, "more by curiosity than by necessity." The records indicate that the patient was a butcher, perhaps the very one recruited as a helper in the transfusionist's earlier experiment.[1] By profession, the man lacked a fear of blood and remained jovial as he marveled at his pulsing veins, presumably ripe with sheep blood. Once the procedure was over, he leaped merrily from the table and flayed the donor lamb in an impressive show of his professional skills. Not one to waste a good animal, he then asked Denis if he might take the lamb home for supper.

Pleased with the results of the experiment, Denis was nonetheless enraged to find the man at the tavern a few hours later, as boisterous as ever—and drunk. In the late seventeenth century, just about every crowded city street in Paris had at least two or three such watering holes. Sporting names such as La Fosse aux Lions (the Lions' Ditch) and Le Berceau (the Cradle), taverns were sites for locals to imbibe, quarrel, meet prostitutes, or just unwind from the fatigue of everyday life in the bustling capital.[2] Paris was heaven for drinkers, and the butcher was a drinker. The transfusionist had paid the butcher in money as well as meals, and now grimaced as he saw how it had been spent. The staggering man slung an arm over Denis' shoulder and, slurring his words, said he had never felt better. When could he and his drinking buddies get another one of these blood experiments? As annoyed as Denis was, his patient's enthusiasm confirmed—of this the transfusionist was certain—the brilliance of his work. Maybe the man's good spirits were the result of the transfusion or maybe it was just the wine that had also flooded his veins. In any event there was indeed cause for celebration: The man was still alive.

On June 25, 1667, Denis sat confidently at his writing desk, dipped his pen in the inkwell, and drafted a letter to Montmor that laid out every detail of his successes. Denis' "Letter Concerning a New Way of Sundry Diseases by Transfusion of Blood" may have seemed intended as a private correspondence between the transfusionist and his patron, but nothing could have been further from the reality. After all, Montmor had attended many of the transfusion experiments himself and knew firsthand the details of these procedures. As soon as the ink was barely dry on the stiff parchment pages, Denis' letter was instead whisked off to a printer's shop on Paris's rue Saint-Jacques and would soon be distributed broadly both in Paris and, of course, across the Channel.

In the letter, Denis claimed to be the first physician to have performed a human blood transfusion. This was more than the English could bear. To their great consternation Jean-Baptiste Denis was an imitator par excellence. His idea to transfuse dogs was inspired directly by the experiments performed at the Royal Society. Indeed, the Frenchman's move to cross-species experiments had been pulled directly from the pages of Boyle and Lower's sixteen-point memo in the *Philosophical Transactions.*

The animosities surrounding blood transfusion were not simply a matter of scientific rivalries; they were one more piece of an increasingly complex political puzzle. The globe was ever-expanding in this age of scientific and cultural exploration, and the ports of each of the major European countries—France, England, Holland, and Spain—teemed with activity as ships set off to stake claims on portions of the world that had only recently been discovered. This quest to dominate the trade routes to and from the New World and Asia exacerbated long-standing tensions among the European nations. Peace, when it could be had, was built around fragile alliances, treaties, and royal marriages. Unrest between two or more of these major political and military powerhouses could set all of Europe on edge and, if not contained, could light the entire continent on fire. It was often on the broad expanses of the high seas that the greatest threats to international relations loomed—and sometimes for the smallest of reasons. Custom had long dictated that two ships of different nationalities should salute each other when passing at sea. This was accomplished by firing a salvo of cannons or briefly lowering colors. But the ever-haughty Louis XIV bristled at naval convention and ordered his admirals and commanders to insist that every foreign ship display submissive homage to the French colors. Not surprisingly other nations resisted mightily—and to such degree that French and English ships, in particular, tried to avoid one another entirely in order to avert certain conflict.[3]

The year 1667 was marked not only by Denis' now-infamous animal-to-human blood trials. It was also a year when international relations had been pushed to their limits. France found itself pitted once again against its Spanish neighbors as the two countries battled over rights to the Spanish Netherlands. Spain had long held control of the seventeen provinces of the Netherlands. In 1648, after eighty years of war, the northern provinces had been granted independence. The southern territories that now comprise Belgium, Luxembourg, and parts of northern France, however, remained under Spanish control. The Sun King was eager to annex this small area between France and the Netherlands to his own empire. And following the death of the Spanish king Philip IV, Louis was convinced that his wife, Marie-Thérèse, the eldest daughter of Philip's first marriage, had full rights to the territory as part of her father's succession. Of course Spain resisted, claiming that those rights were to be transferred instead to the children of Philip's second marriage, who were still minors.

The "War of Devolution," as it was called, became still more complicated when the independent Dutch provinces found themselves at war with England. The Netherlands had experienced phenomenal success in the spice colonies and had quickly established itself as a major economic player in late-seventeenth-century Europe. As allies of the Dutch, the French were drawn back into battle with the English. Yet Louis XIV was urged by his advisers to be cautious in his dealings with Charles II. Rumors were afloat that the English king was more than ready to align himself with Spain. This would complicate French designs on the Low Countries and launch France into what could be a decades-long war on multiple fronts. Louis XIV offered naval assistance to the Dutch, but he ordered his forces to avoid at all costs engaging directly with the English.

While the king may have shown uncharacteristic restraint, Jean-Baptiste Denis was not so politic in his dealings with the English. If Denis had found few friends among his fellow countrymen, he would soon find still even fewer among the English. The transfusionist had declared his own one-man war against England's Royal Society. And the influential secretary of the society, Henry Oldenburg, would not quickly forgive Denis his transgressions.

Oldenburg spent his days reading, organizing, translating, and responding to the letters that flooded his mailbox. The amount of work required was, as he complained to Boyle, simply overwhelming: "I am sure no man imagines what store of papers and writings pass to and from me in a week from time to time, [of] which I rid myself without any assistance. I have no less at present than thirty correspondents, partly domestic, partly foreign. Many of them I am not only to write to, but also to do business for, which requires much time to inquire after such particulars and dispatch such business."[4]

Oldenburg's work was a true labor of love. It had to be, for his was certainly not a position that offered much financial reward. The job of secretary of the Royal Society was not salaried; in fact Oldenburg received only rare reimbursement for his endless paper, ink, and postage expenses. He had earned some money by doing private translations for Royal Society colleagues, especially Boyle, and by writing news-filled letters to virtuosi outside London, but he frequently lamented his pennilessness and was always seeking new sources of revenue. Plague, fire, and unscrupulous publishers had created nearly insurmountable obstacles for Oldenburg. To complicate his financial situation still more, his wife of just a year and a half died in the months following the first issue of the *Philosophical Transactions.* The widower was obliged to commit the bulk of his wife's modest dowry to her funeral. While the publication had been successful, it hardly brought in the money that he so desperately needed. "What was hoped," Oldenburg lamented bitterly to Boyle, "might have brought me in about £150 per annum . . . . will now scarce amount to £50."[5] He later estimated that his work as editor brought in even less, just £40. Given the extraordinary number of letters Oldenburg wrote and received, it is not easy to understand how he shouldered what would have been exorbitant postal expenses. In early Europe it was the responsibility of the recipient to pay all postage expenses at the time of delivery—and the expenses were not insignificant. A single sheet of paper traveling just eighty miles within England could cost upwards of twopence. Yet a large percentage of Oldenburg's letters came from the Continent, which would have likely quadrupled the fees, or more.[6]

It was well known across England that Oldenburg actually corresponded frequently with Dutch and French scholars at the Paris Academy of Sciences: Auzout, Petit, and Huygens. He had also recently begun exchanging letters with Henri Justel, Louis XIV's personal secretary, who shared the latest scuttlebutt from the French court. And now, at the height of tensions with the French and their allies the Dutch, news of Denis' experiments had been arriving with some regularity at Oldenburg's home. On June 20, 1667, Oldenburg received the letter that would set off a series of fireworks in the English scientific community—fireworks that would not be extinguished until well into the next fall.[7]

In the weeks and months that followed his first animal-to-human experiments, Denis was in the thick of launching a major self-publicity campaign. In his announcements Denis conveniently neglected to acknowledge his debt to the English. He mentioned neither Harvey or Wren nor Lower or Boyle, on whose work his own experiment had so obviously relied. In scientific circles it had long been the tradition to recognize, however briefly, the forerunners to a particular theory or discovery before launching into a celebratory description of one's own successes. But Denis had cast aside the work of the Englishmen who were now well-known across Europe for their blood studies. Instead he had begun his letter with a proclamation that blood transfusion was first proposed in France. In his opening comments to Montmor, he wrote:

> Sir,
> The project of causing the Blood of a healthy animal to pass into the veins of one diseased having been conceived about ten years ago, in the illustrious Society of Virtuosi which assembles at your house; and your goodness have received M. Emmerez & myself, very favorably at such times as we have presum'd to entertain you either with discourse concerning it, or the sight of some not inconsiderable effects of it: you will not think it strange that I now take the liberty of troubling you with this letter, and design to inform you fully of what pursuances and successes we have made in this operation; wherein you are justly entitled to a greater share than any other, considering that it was first spoken of in your Academy.[8]

The Frenchman credited the idea of blood transfusion to a man few had heard of—a man of whom little historical trace is left. According to Denis' account, an unknown Benedictine monk, Dom Robert Desgabets, first proposed the idea of "blood transfer" (*communication du sang*) to the Montmorians in July 1658. Desgabets suggested that donor blood could be collected in a leather pouch and then poured into a silver pipe. One end of the pipe would be large, like a funnel, to receive the blood. The other end would be thin and narrow, so that it could penetrate the vein of an animal or a human. No trials were performed using this method, but Desgabets' ideas—asserted Denis—were proof enough that blood transfusion was French.

Denis' claims were all the more surprising because, in 1658, Montmor's academy had shown almost no interest in medical topics. Instead the Montmorians had thrown themselves headlong into astronomy and had been eagerly awaiting news of Huygens's studies of Saturn. With the exception of Denis' solitary account, no extant historical documents confirm Desgabets' alleged presentation at the Montmor Academy. What is more, had Desgabets actually presented his ideas at the academy, Denis would have been unlikely to have heard them firsthand. Little more than twenty-two years old at the time, he would not yet have finished medical school, and as a young man of lower-class origins, he would hardly have been invited to meetings at Montmor's estate. Denis' self-assured assertions were, then, no more than wishful hearsay or, worse, complete fabrications.

John Wallis, a fellow of the Royal Society, had anticipated Denis' arrogant claims. He had long been worried that the society was insufficiently aggressive in its efforts to take full ownership of its discoveries. Wallis had shared his concerns with Henry Oldenburg as early as March 1667, nearly three months before Denis' scandalous letter. Wallis noted the publicity that had surrounded Denis' thievery—which he called the "French operation in imitation"—and told Oldenburg that he could "only wish that those of our own Nation were a little more forward than I find them generally to be in timely publishing their own

Discoveries, & not let strangers reap the glory of what those amongst ourselves are the Authors."[9] Wallis's warnings were certainly prescient. Now Denis' letter had just upstaged English claims to dominance in the blood wars. Wallis's message was a clear critique of Oldenburg's work at the Royal Society. The tasks Wallis described fell squarely on Oldenburg's shoulders as both secretary of the society and editor of its *Philosophical Transactions.* If discoveries by Royal Society members were not recorded appropriately or announced in a way that glorified the society's intellectual and scientific endeavors, it was Oldenburg who would have to answer for it.

The timing of Denis' letter could not have been worse for Oldenburg politically. Xenophobic tensions usually bubbled over in England in the wake of disaster, and following England's recent disgrace at the Battle of Medway, animosities toward foreigners had hit a new high. Earlier that month, on the morning of June 6, a dense fog had cloaked the English coast near the Isle of Sheppey. Quietly and undetected, nearly one hundred Dutch ships entered the well-protected estuary of the Thames and sailed into the nearby river Medway. By the time the fog cleared and the alarms were sounded, it was too late. The Dutch fleet continued upriver and toward the shipyards of the huge naval base at Chatham, where it captured the one-hundred-gun *Royal Charles.* The Dutch gloated while the British navy's largest and best-equipped battleship was towed back to Rotterdam. As Pepys noted shortly after the defeat, "The Dutch do mightly insult of their victory, and they have good reason."[10] Lord Arlington, Charles II's powerful secretary of state, scrambled to deflect any blame for the stunning defeat from himself. The commissioner of the Royal Navy at Chatham, Peter Pett, was made the official scapegoat for the Medway disaster; he was promptly imprisoned in the Tower of London. Immediately following the attack Arlington set out to expose other "traitors" who might have helped smooth the way for England's disgrace.

As the violence inflicted on the French and Dutch following the Great Fire had shown, foreigners were always a prime target in seventeenth-century England when both the government and the populace sought to exact vengeance for their losses. The German-born Oldenburg knew he was at risk during this moment of high international tension that followed the battle at Medway. He may have integrated himself seamlessly into English society, but he would never be considered fully English. And he was well aware that his prodigious correspondence, combined with his mastery of more than seven languages, could leave him vulnerable to intense government scrutiny. Oldenburg was now at the top of Arlington's list of traitors.

Given his frequent communication with colleagues on the Continent, Oldenburg had been wise enough to protect himself from royal spies and postal censors. Established in 1635, the post office served as much as a mechanism of domestic surveillance as a means to ensure the timely delivery of packages and letters.[11] Chief among the censors was the inventor Samuel Morland, who occupied a secret room adjacent to the General Letter Office. Morland had even taken one of Arlington's own letters and made several copies, apparently returning the original unopened—proving, as the French ambassador Comminges noted, that the "English have tricks to open letters more skillfully than anywhere in the world."[12]

Royal censorship posed a threat to privacy that the wealthy could not risk; for this reason they paid private couriers who creatively disguised their letters and packages to avoid detection. "Several letters I carried to and brought from France," wrote one courier, "were made up as the mould of a button, and so work'd over with silk, or silver, or worn on my clothes. Others I brought over in the pipes of keys."[13] Some especially cautious writers bound their letters in books or wrote in code or with invisible ink. Penning their words in a clear solution of vinegar mixed with lead oxide, they would overwrite them with a less-secretive message in visible ink. The recipient would use arsenic trisulfide and limewater to dilute the visible ink, turning the invisible script gray and legible. Human urine was another liquid sometimes used for invisible ink; the writers would dip the end of the quill in the urine and lightly trace their words onto the paper. Letters could then be held over the steam of "a compound of several spirits, metals and sulphur boyl'd together and made liquid" to reveal their contents.[14]

Oldenburg did not have the resources to employ such elaborate measures. Instead he approached fellow Royal Society member and Keeper of State Papers, Joseph Williamson. Williamson had a quiet reputation for the "illicit side of the Post Office's activities."[15] The two men decided that all of Oldenburg's incoming correspondence would be addressed to a pseudonym, "Mr. Grubendol, London." Williamson retrieved the letters and gave them to Oldenburg unopened. In return Oldenburg agreed to provide Williamson with excerpts of any political news that the letters contained. For this, Oldenburg received reduced mailing fees, and Williamson could stay apprised of activities abroad.[16] But it soon became clear that Williamson's "protections" could only go so far.

Royal guards surrounded Oldenburg's modest home in Pall Mall. We can only imagine that Oldenburg was filled with confusion and fear as he was whisked away from his home without warning. He was taken by coach to the banks of the Thames and from there likely escorted onto a boat that would move him to the Tower. Water transfer was more secure than transportation through the streets and over the London Bridge, where carriages could be easily overtaken to allow prisoners a chance to escape the terrifying fate that awaited them. Prisoners charged with treason entered on the banks of the Thames through the "Traitors' Gate," where more famous figures like Anne Boleyn, Sir Thomas More, and Sir Walter Raleigh had entered. Two large barred gates slowly creaked open to reveal several dour-faced guards who awaited their next prisoner. The guards led Oldenburg through the gateway of the menacing "Bloody Tower" and to his dark and modest cell.

The charges against Oldenburg were vague, accusing him simply of "dangerous desseins and practices." But for contemporaries such as the diarist Samuel Pepys, there was little doubt that Oldenburg's connections with the French had everything to do with Lord Arlington's suspicions: "Mr. Oldenburg, our Secretary at Gresham College, is put in the Tower for writing news to a virtuoso in France, with whom he constantly corresponds in philosophical matters; which makes it very unsafe at this time to write, or almost do anything."[17] Historians have been unable to pinpoint with certainty the exact source of Arlington's suspicions. Yet Oldenburg received Denis' letter the very day of his arrest.[18] While we cannot know if it was the precise cause of his arrest, it is nonetheless certain that Denis' letter, with its outrageous claims of French superiority, did little to help Oldenburg's case.

Two weeks later, in Oldenburg's absence, the president of the Royal Society, John Wilkins, summarized the contents of Denis' letters to his colleagues at the society's July 4 meeting. The report was greeted with outrage. As Royal Society fellow Timothy Clarke later fumed, "I am not so clear why that learned Frenchman disputes so vigorously and so warmly over the origin of blood transfusion." With barrister-like precision, Clarke refuted in writing Denis' claims that the French were first to imagine blood transfusion, and reviewed the history of English blood experiments. Clarke cited John Aubrey, who documented Francis Potter's suggestions in 1639 that transfusion would be an ideal way to test Harvey's ideas on blood circulation. He confirmed that later, in 1653, Potter had apparently collected the blood of one animal in a dish and tried transfusing it into another by using ivory tubes and quills. The procedure failed, ostensibly because of the time lapse between collection and transfusion, which caused the donor blood to clot.[19] Clarke then moved to Christopher Wren, who "first thought of (and performed at Oxford) the injection of various liquors into the mass of the blood of living animals." Clarke further claimed that, in the following year, he himself had injected "waters, various kinds of beer, milk and whey, broths, wines, alcohol, and the blood of different animals" into dogs. Clarke concluded his arguments by describing Richard Lower's canine experiments in 1666, asserting that this should be sufficient evidence that "the honor for this invention—if it deserves any—should be awarded to the English rather than the French."[20]

Now, denied pen and paper in his oppressive prison in the Tower of London, Oldenburg had no way of communicating that he was innocent. A single letter was delivered to his cell a few weeks after his imprisonment. It was from Williamson, who urged Oldenburg to remain patient—he would soon be released. While still fearful that he might not leave the Tower alive, at least he now had a piece of paper. Oldenburg begged his jailers for ink and a pen. He received this "particular favor"—for which he paid handsomely, no doubt—and wrote on the back of Williamson's letter the urgent pleas of a man who believed that his days were numbered.

> S^r I thank you for your friendly letter: I pray, continue your kindness, as far as you may, and, when you see it seasonable, present my very humble service to my Lord Arlington, telling him that I hope his Lord will have experience in time, when this present misunderstanding shall be rectified, of my integrity and of my zeal to serve his Majesty, the English nation, and himself to the utmost of my power. Meanwhile, I beseech you, be pleased, when you find it seasonable to cast in a word of the narrowness of my fortune for to lie long in so chargeable a place as the Tower is. What you shall think fit to send to me of the papers that are come to your hands for me, will be a welcome diversion too.
>
> Sir, Your obliged and humble Servant, H. Oldenburg.[21]

Oldenburg was right to be concerned about the "narrowness of his fortune" while in the Tower. Early prison protocol required prisoners to pay their own room and board. The choice of lodging in England's most notorious prison was dictated by the amount that a "guest" could pay. By law, jailers could also charge extra for *sauvitas* (gentle keeping). And depending on the payment received, accommodations could range from comfortable to squalid—that is, from a spacious room with a bed, a desk, and a view to a windowless cell shared with several other men and straw on the floor for a bed. Even exonerated of charges, prisoners could and did remain in jail indefinitely if they were unable to clear the debts incurred during their stay. And debts could accrue very fast.

There is no evidence that Williamson replied; Oldenburg paced alone in his dark cell, anxious and deprived of writing material. About two weeks after his correspondence with Williamson, Oldenburg received a visitor who provided him with some glimmer of hope that he might leave the Tower alive. Oldenburg was able to persuade the visitor, whose name remains unknown to us, to write a letter to the bishop of Salisbury to intervene on his behalf. Oldenburg was unnerved by the fact that no official charges had been levied against him but still he had been left to languish in one of London's most notorious jails. "I am not guilty of anything," Oldenburg wrote via his acquaintance, "and all who know me well can attest my love, concern and zeal for the King's and the kingdom's interest and prosperity. Besides, I have employed even my correspondencing to give advertisement to the Court, such as I thought might be useful to England. . . . And now I beseech your Lordship that you would please to take all the opportunities you can to represent me to his Majesty and to my Lord Arlington, and with all to engage such of your noble friends, as are in the King's favor."[22] Oldenburg's pleas fell on deaf ears. An exercise in futility, the letter never made it to the bishop; it was confiscated from the visitor as he left the Tower.[23]

One month after his arrest Oldenburg's dire situation took another turn for the worse—and once again, Jean-Baptiste Denis was at the center of the problem. Since its beginnings in 1665, the *Philosophical Transactions* had been considered nearly synonymous with its editor, Oldenburg. Clearly he was in no position to arrange for the publication of the journal's next issue. Yet, by some mystery, a new issue did appear on newsstands and in hawkers' hands on July 22. The issue featured a full translation of Denis' controversial "Letter Concerning a New Way of Sundry Diseases by Transfusion of Blood"—including the opening portions in which the transfusionist credited the French with originating the procedure. There was something suspicious about this issue from the beginning: It carried the correct date and was paginated consistently with the previous issue, yet it lacked the standard *Philosophical Transactions* header.

Historians have speculated that, during Oldenburg's absence,

Royal Society president John Wilkins had arranged for the publication of this next issue of the *Philosophical Transactions.*[24] Yet Denis' letter had created an uproar among Royal Society members, so it is not entirely clear why Wilkins would have reprinted it without commentary or disclaimer. Another, perhaps more likely possibility is that the issue was a counterfeit created for financial gain or as an effort to construct additional evidence against Oldenburg and the treason charges he faced. There is no archival proof for either explanation, and the true circumstances behind this anomalous issue will forever remain a mystery. In any case, one thing is certain. Oldenburg was enraged when he discovered that the *Philosophical Transactions* had been published without his approval—and was doubly anguished to learn that his cherished journal had been used to circulate Denis' lies.

The Anglo-Dutch hostilities ended on July 31 with the signing of the Treaty of Breda. Animosities and suspicions subsided, and soon many of the men who had been charged with treason were exonerated. Two months after his arrest a grateful Oldenburg was released from the Tower. He left immediately for the countryside in order to recover from his hair-raising ordeal. "I was so stifled by the prison air," Oldenburg wrote to Boyle shortly after his release, "that as soon as I had my enlargement from the Tower I widened it, and took it from London into the country to fan myself for some days in the good air of Crayford in Kent." Still, he worried deeply about his reputation and knew that he needed desperately to prove his allegiance to his adopted country. "My late misfortune I fear will much prejudice me," he wrote, "many persons, unacquainted with me, and hearing me to be a stranger . . . spread it over London, and made others have no good opinion of me. . . . I hope I shall live fully to satisfy his majesty, and all honest Englishmen of my integrity and of my real zeal to spend the remainder of my life doing faithful service to the nation, to the very utmost of my abilities."[25]

When Oldenburg returned to London, he still had much unfinished business when it came to Denis and the havoc the French transfusionist had caused in the Royal Society and for Oldenburg personally. Oldenburg also had to smooth many ruffled feathers—starting with Richard Lower's. Lower wasted no time in arriving unannounced at Oldenburg's home in order to lodge a complaint. The normally stout Oldenburg looked gaunt; he had not eaten well in the Tower and had only recently regained his appetite. Weakened and pale, Oldenburg approached the enraged Lower with deference and begged him to believe that he had had no role in the publication of the spurious issue of the *Philosophical Transactions*. If he had indeed been the one to publish Denis' letter, a nervous Oldenburg explained, he would have most certainly added an "Animadversion" that refuted Denis' outrageous claims. Oldenburg promised Lower that the next issue of the *Philosophical Transactions* would make it abundantly clear that Denis was a liar.[26]

On September 23 Oldenburg made good on his promise. He set the record straight in a new issue of the *Philosophical Transactions*, one that covered all that had gone unpublished while he had been imprisoned in the Tower. "It is notorious," Oldenburg wrote, "that [transfusion] had its birth first of all in England; some ingenious persons of the Royal Society having first started it there several years ago and that dexterous Anatomist Lower reduced it into practice, both by contriving a method for the Operation, and by successfully executing the same."[27] Oldenburg reminded readers that earlier issues of his *Transactions* had documented this abundantly.

A month later the secretary of the society dedicated another full issue of the *Philosophical Transactions* to exposing yet again the arrogance of the French transfusionist. In particular he charged Denis with a shocking disregard for the safety of his patients. If the English were moving more slowly than the French—or more particularly, Denis—Oldenburg emphasized that it was because his fellow countrymen were practicing an abundance of caution. "They [the French] must give us leave to inform them of this Truth, that the Philosophers in England would have practiced long ago upon Man, if they had not been so tender in hazarding the life of Man (which they take so much pain to preserve and relieve."[28]

Oldenburg knew that his refutation of Denis' claims, no matter how spirited, would likely be insufficient for him to regain all that had been lost. It must have come as some consolation, however, that one of the highest-placed members of the French court had also agreed that Denis had been out of line. Even the most prudent of English censors could not have taken exception to the letter that Louis XIV's secretary, Henri Justel, sent Oldenburg later in the fall: "I must admit that [Denis] was too credulous in accepting the statement of those who said that transfusion was discovered in France rather than in England. I have told him he ought to inform himself more carefully than he has done. All honorable men agree with your opinion."[29]

Denis had no intention of heeding Justel's warnings. The French transfusionist was more than aware of the furor that was directed against him from all corners, both at home and abroad. Truth be told, he reveled in it. A man of humble birth, he had beaten the odds and had come much further than anyone—perhaps even he—would have thought possible. He was not going to stop now.

Chapter 12

BEDLAM

It would be a long time before Oldenburg forgot the personal and financial toll that the transfusion controversy had taken on him. Denis may have stretched the truth about transfusion's origins, but neither Oldenburg nor the English scientific community could deny that the French had won the race for the first human blood transfusion. Now the fellows of the Royal Society wondered if they had been too cautious about experimenting with the procedure in humans. It was no secret that Edmund King and Richard Lower had been ready for nearly six months to begin human experiments. But they had been waiting for the "removal of some considerations of a moral nature"—that is, a stronger consensus in the Royal Society that such trials could, and should, be attempted despite their evident dangers. The "ingenious" Doctor King documented this timeline in a letter he sent to Oldenburg in the weeks following the secretary's release from the Tower; his clear intention was to see it printed in the *Philosophical Transactions.* Oldenburg was happy to oblige. If anything his time in prison had made him an even more dedicated advocate of the English scientific cause. He wanted there to be no doubt of his loyalties to his adopted country and to his colleagues at the Royal Society. King's letter was published in full on October 21, 1667:

> Sir,
> The method of transfusing blood you have seen practiced, with facility enough, from beast to beast; and we have things in a readiness to transfuse blood from the artery of a lamb, kid, or what other animal be thought proper, into the vein of a man. We have been ready for this experiment for six months, and wait for good opportunities, and the removal of some considerations of a Moral Nature. I gave you a view, you may remember, a good while ago, of the Instruments I think very proper for the Experiment, which are only a silver tube, with a silver stopper somewhat blunted at one end, and flattened at the other for the conveniency of handling, used already on beasts with good success.[1]

While the English hesitated, Denis had pounced—and taken all the credit. Trying not to look back at lost opportunities, Oldenburg was now more certain than ever that the English needed to launch boldly into human trials once and for all. One month after his "enlargement" from the Tower, Oldenburg stood in front of his colleagues and made a motion before the entire Royal Society that blood transfusion "be prosecuted and considered, in order to try it with safety upon men."[2] His proposal helped mend fences with critics who had criticized him for not doing enough to promote the English cause, and it was accepted without hesitation. The English were now back in the game.

Refusing to be scooped again, the entire Royal Society worked concertedly to prepare for human transfusion. In haste Richard Lower—the English "father" of blood transfusion—was appointed an official fellow of the Royal Society. The society rented a room near their regular meeting place where the anatomist could perform his experiments in collaboration with King. Lower's laboratory space was conveniently situated along the Thames and offered a view of the river; but, more important, it provided an easy means to discard the carcasses and entrails of the animals on which they would experiment. At the next meeting of the society, King read aloud his detailed "method of transfusing blood into a man" and requested that it be registered in the official society record. The only thing that remained now was finding a willing patient.

Jean-Baptiste Denis' animal-to-human trials had been performed first on a boy suffering from an untreatable fever; his second patient was a healthy but drunken middle-aged butcher. For the English to distinguish themselves in the experiment, they would have to select a subject who was very different from those in the French trials. Yet, their first patient would also have to be someone who would survive the blood transfusion and, better still, show a marked improvement in health because of it. At one of the society's next meetings, George Ent, a respected anatomist and close friend of the late William Harvey, proposed that the society try the experiment on "some mad person in the hospital of Bethlem."

Bethlem, or Bethlehem, Hospital was founded in 1247 by the religious order of Saint Mary of Bethlehem. Situated just east of Bishopsgate and outside the walls of London, it served initially as a hospice for the ill and poor of the community. In 1547, however, Henry VIII claimed the hospital on behalf of the government and officially declared it London's home for "melancolicks" and the "troubled in the mind." Now, a century and a half later, the hospital was overflowing with patients, pestilence, and the never-ending din of human misery. Since the Middle Ages, Bethlem had also been called Bedlam; and the asylum was the very incarnation of the chaos with which its name would become synonymous. Bedlam was comprised of just a few small stone buildings, a tiny church, and a garden. Conditions were lamentable, if not horrific. The hospital was perennially understaffed. Raw sewage lay stinking both in- and outside the living quarters, which were crammed with filthy, suffering men and women. The Great Fire of London the year before had spared the hospital; and the frenzy of new construction throughout the city simply underscored Bedlam's dilapidated and brutal state of affairs. Shivering under leaking roofs, the inhabitants of Bedlam were tormented twofold: by their troubled minds and their hellish living conditions.

If the "horrors of Bedlam" were undeniable, they were doubly so for the most agitated and menacing of patients, who were often chained to the walls. There is little doubt that early English society was violent, even when compared with modern standards. Hitting and flogging were common in the general populace, especially toward persons of subordinate status, like women and children. But Bedlam's chains and unfettered violence were more a manifestation of the fears of those who came face-to-face with uncontrollable madness than they were about real hopes of curing the asylum's condemned. Extreme madness was unsettling and needed to be beaten back at all costs. Meanwhile, those who suffered only mild mental illness did not evoke such intense reactions and were more easily tolerated. In fact, as a means to ease overcrowding, the less "extravagant" Bedlamites—or "bedlam-beggars" and "Tom O'Bedlamers," as they were also called—were released and given license to beg. They were recognizable by a tinplate badge that they wore on one of their arms, which allowed officials to return the mentally ill to Bedlam, should their demeanor move from mild lunacy to full-out insanity. [3]

In early Europe human experiments were rare but not unheard of. The noted chemist Robert Boyle had tested laundanum, an opium-based tincture, on one of his servants, who suffered regular nosebleeds. In 1650 he had also paid a man to let himself be bitten repeatedly by snakes in order to test the hypothesis that a hot iron applied near the bite would neutralize the poison. The cure worked, and the volunteer earned a living by repeating the experiment for curious onlookers.[4]

Bedlam collected men and women along a broad continuum of health and illness—and as such, promised a wealth of poor souls on which blood transfusion could be tried. In short order a committee comprised of Richard Lower, Richard King, Robert Hooke, and Thomas Coxe—all men with hands-on experience in canine blood experiments—was charged with visiting the asylum. Their task: to pick just the right subject for their experiments.[5]

The men had agreed that the cooling effects of blood transfusion could be very promising treatment for "extravagant" minds. At the time, humoral imbalances were still understood to lie at the root of madness. Each of the humors was associated with specific qualities and was sensitive to the influences of the seasons. Blood was considered hot and moist and was most abundant in the spring (which gives new meaning to the term "spring fever"). People who were sanguine by nature were seen as being high-energy, warmhearted, and easily prone to fits of anger. In contrast, black bile was cold and dry, and was most prolific in autumn. Melancholics were milder-mannered and had less energy than sanguine people. Yet their state could range from despondent to suicidal when they suffered from an overabundance of black bile in their bodies. The humors at once influenced and were influenced by human emotion. Heartbreak, stress, and anger could increase body temperature, which would create nox-

FIGURE 19: Life at Bedlam depicted by William Hogarth in *A Rake's Progress* (1735). Some historians have suggested, and perhaps rightly so, that Hogarth may have exaggerated somewhat the conditions of Bethlem. Yet, it is worth noting that the Bedlam at the time of this painting was the more modern and spacious one built by Robert Hooke in 1670. Hogarth's images may just have been—regrettably—right on target for the period in which the Royal Society visited the hospital in search for the ideal patient.

ious vapors in the body that would rise to the brain and cause mental disturbances. The cures for mental illness were, then, very similar to those of any illness caused by humoral imbalance. In these cases bleedings were performed on the forehead or even on the hemorrhoidal veins in order to draw blood down and away from the brain. For "melancholy and mopish people," cooling mixtures of lapis lazuli, hellebore, cloves, or licorice powder were regularly infused in white wine and borage, again as an attempt to cool the humors and calm the mind.[6]

Another more invasive procedure was often performed when humoral therapeutics proved ineffective. Doctors speculated in those cases that a foreign object lodged deep in the brain was the actual cause for the patient's odd behavior. The operation, performed with some regularity since the Middle Ages, involved boring a hole into the skull with a hand-cranked circular drill. Once the barber-surgeon made a sufficiently large hole in the patient's cranium, he would then probe the patient's brain in search of a pea-size "stone."

For a Bedlamer skull-drilling might have seemed preferable to the procedure the Royal Society was imagining for its next patient. The fellows contacted Doctor Allen, manager of Bedlam, to see if he could recommend a patient well suited for an experimental blood transfusion. While we have no details about the conversation that may have transpired between the fellows and the doctor, there is little doubt that Allen refused to go along with the plan. Hooke, Clarke, Lower, and King met personally with the Bedlam head to persuade him to change his mind, without success. Given the inhumane conditions at the asylum, it does indeed seem odd that the physician would have refused on the grounds of patient safety. Yet blood transfusion was still so new and its effects still so dubious that Allen may have refused on ethical grounds. In the absence of historical documents regarding the details of Allen's meetings with the Royal Society fellows, we are left only to speculate.

The transfusionists would have to find another way to locate a suitable subject for their trial. If they were not able to procure a patient directly from Bedlam, they would look for someone who roamed freely through London but who had not yet been

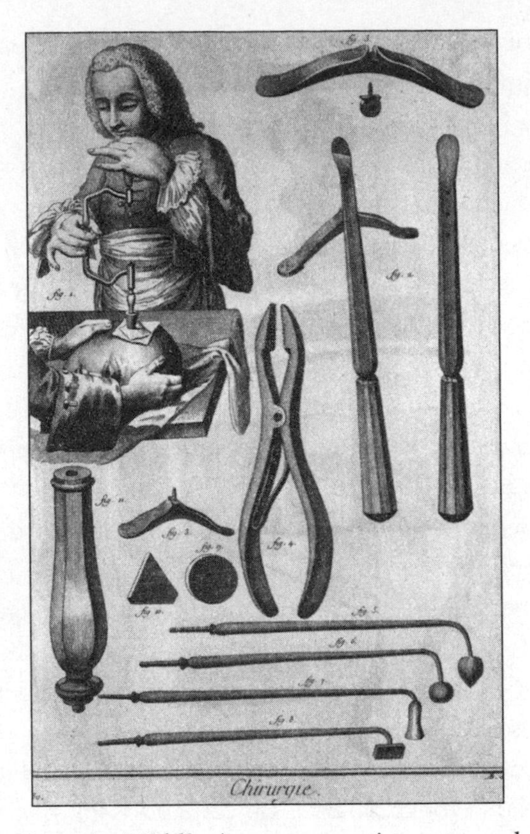

FIGURE 20: Since the Middle Ages, trepanning was used as a way to relieve the symptoms of mental illness. This illustration shows the various tools that were used in the seventeenth and eighteenth centuries for the procedure. *Encyclopédie* (1772).

committed to the hospital. It would not be a hard task. In the busy capital there were as many, if not more, mildly deranged people on the streets as there were within the walls of a single madhouse.

The Royal Society committee soon learned from fellow member John Wilkins about a possible candidate. The thirty-two-year-old Arthur Coga attended Wilkins's church. Coga had studied at

Pembroke College, Cambridge, where his brother would eventually become a schoolmaster. Obviously well-educated and from a respectable family, Coga preferred to speak Latin, insisting on using it for any possible occasion. And that was not the only curious thing about him. There was something a little off about the man's behavior, but no one could put a finger on it. The only diagnosis that history has left us comes from Richard King, who explained simply to Boyle that Coga's "brain is sometimes a little too warm." Oldenburg also acknowledged that Coga was "look't upon as a very freakish and extravagant man . . . an indigent person." His assessment was confirmed by Wilkins, who told Pepys during a drinking session in a London tavern that Coga was "a little frantic . . . a poor and a debauched man."[7] The prevailing logic was that transfusion would help cure Coga of his illness by replacing his overheated blood with other, cooler blood. As such, he was the perfect candidate for the experiment that would put England back into the blood wars.

Shortly before eleven o'clock one brisk morning in late November 1667, Edmund King arrived at Lower's laboratory near Arundel House. News of the scheduled experiment had traveled through the society, and the surgeon was forced to push his way through a thicket of more than forty spectators who had come to observe the transfusion. The lengthy guest list included several physicians, members of Parliament, and even a bishop. Before these witnesses King and Lower began by opening the artery of a sheep. They placed a small pan to catch the blood that flowed out of the animal while they worked to insert a narrow silver tube into the blood vessel. They let dark red liquid from the animal's artery flow into a dish and measured it; from this the surgeons calculated a flow rate of about twelve ounces of blood per minute. Then they swiftly capped the tube with a silver stopper. Some blood continued to seep from the wound, but for the most part, the stopper mechanism kept the lamb's blood where it needed to be for now: in the beast.

Next they turned to Coga, who was admiring the "florid arterial blood" resting in the porringer. Mesmerized, the man dipped a knife into the pan and brought it to his mouth. He liked what he tasted. Finding it "of good relish," Coga stretched his arm eagerly toward Lower and King.[8] The surgeons used a fine-bladed lancet to open one of the man's veins. They let about seven ounces of blood, making room for the quantity of lamb's blood with which they intended to replace it. One of the surgeons gripped Coga's arm just below the incision to reduce blood flow until the other could slip another small stoppered tube upward into the vein. Nodding to each other, they removed the stopper from each of the tubes in unison and linked them with a series of thin quills. The room fell silent as all eyes moved quickly from beast to quills to man and back. Nearly a full minute had passed, and there was no sign that any blood was moving out of the lamb's artery and into the quills. Lower and King began to worry that blood was clotting in the stoppered tube. They waited anxiously, hoping that the blood would begin to flow. Without warning and to the surgeons' great relief, it did. The red fluid pushed its way through the quills and "ran freely into the Man's vein for the space of two minutes at least." They removed the quills from the tube in the lamb's artery and disconnected the tube from their patient's arm.

By every description Coga made "not the least complaint" during the procedure. The experimenters asked the man several times how he was feeling. In Denis' experiments both the boy and the butcher had complained of heat at the transfusion site. Coga showed no such signs, to the smug satisfaction of Lower and King. King speculated that, during the nervous moments as they waited for the blood to begin flowing through the quills, the fluid had cooled and "come in a temper very agreeable to venal blood." To assure spectators that blood had indeed moved from the animal and into Coga, they did not replace the stopper. Instead they bled the sheep dry in a "very free stream." King collected some of the blood and made a few quick calculations. Their experiences showed that donor blood flowed more slowly in the second minute of a transfusion than it did in the first; they had also bled about seven ounces of blood from the man before initiating the transfusions. Taking these variables into account, King estimated that they had infused about ten or eleven ounces of sheep's blood into Coga.

When it was all over the spectators fawned over Coga, who looked "well and merry." He was an extrovert who loved attention almost as much as he loved a stiff drink. While King stitched him up, he enjoyed a glass of wormwood wine and regaled the crowd with stories in Latin, English, and at times an incomprehensible mix of both. A second glass of wine and a pipeful of good tobacco later, Coga answered a slew of questions about being the first Englishman to have veins full of animal blood. When asked why he was happy with the choice of lamb's blood, the playful Coga smiled and replied without missing a beat: *Sanguis ovis symbolicam quamdam facultatem habet cum sanguine Christi; quia Chistus est agnus Dei* (The blood of sheep has some symbolic association with the blood of Christ, because Christ is the lamb of God). The room erupted in boisterous tavernlike laughter, which transformed into cheers after King checked the man's pulse and announced with pride that it was much "stronger and fuller" than before the transfusion.[9] England's sacrificial "lamb" had survived and even thrived.

Coga returned home a few hours later. His appetite was good and—an important detail for early medical diagnoses, which emphasized urine and excreta—the man had "three or four stools as he used to have before."[10] But by early evening, the high of the transfusion had given way to fatigue that went beyond what Coga's caregivers said was normal for him following a simple bloodletting. He crawled into his bed and slept well, despite an uncontrollable and profuse sweating that lasted for several hours, no doubt the result of a mild reaction following the transfusion of incompatible blood. The next morning Henry Oldenburg and John Wilkins went to the man's home to check on his progress. Coga was awake but was still lounging in bed. The secretary of the Royal Society marveled that Coga, who had once been "lookt upon as a very freakish and extravagant man," seemed very composed and more like the well-educated person that he was.[11] Immediately following the procedure Coga had begged King and Lower to repeat the experiment on him in a few days. They demurred, deciding instead to take a bit more time to review the results of the first trial. Now in the company of Oldenburg, Coga wanted it recorded that he stood ready to serve as a volunteer for another transfusion—provided, of course, that he could expect another payment and some more wormwood wine.

The advantage of using Coga was that he was an educated man. Despite his tendency to disordered ravings, the Royal Society fellows felt sure that he was competent enough to give an accurate account of "what alteration, if any, he doth find in himself, and so may be useful." A week after his initial transfusion, Coga returned to the Royal Society and was put on show for the fellows. Pepys marveled at the man's condition. "He speaks well," the diarist wrote later that evening, "and did this day give the Society a relation of it in Latin, saying that he finds himself much better since." Coga may have felt like a "new man," but Pepys still wondered about the patient's mental state: "He is cracked a little in his head, though he speaks very reasonably and very well."[12]

Oldenburg was euphoric. There was now no doubt that the English had demonstrated their dominance in the blood race. He announced it with pride in letters to the Continent. Oldenburg's need to restore his own reputation—and that of the English more generally—following the Denis fiasco was never far from the surface. His letter to René-François de Sluse, a French-speaking philosopher in Liège, amply demonstrated this. "I cannot conceal from you," wrote Oldenburg, "that our Society has so far succeeded with that experiment of transfusing blood from one animal into another (which they had previously attempted with good fortune on beasts a good many months ago, and which the French have successfully imitated) and that, a few days ago, they performed it upon a man, not without good results." Oldenburg remained optimistic as well that transfusion would prove "most beneficial to humanity, because of the large losses of blood incurred in blood-letting, as in the treatment of frenzy and many other diseases" and promised that he would keep his correspondents abreast of "its further fortunes" in England.[13]

It would not be long before Oldenburg had more news about English successes. The society indulged Coga's requests for another transfusion on December 12, 1667, again paying him twenty shillings for his trouble. Another noisy and excited crowd congregated at Arundel House. Using the same procedure as before, King bled what he guessed was about eight ounces of blood from Coga and replaced it with what he announced was about fourteen ounces of sheep's blood.[14] Many in the crowd shouted that they did not believe that such a great quantity could have traveled into the man's body, insisting that, next time, better efforts should be made to weigh both animal and man before and after the procedure.

The crowd's skepticism was actually right on target. There is a strong possibility that little to no blood was actually transferred.

The distance between the donor animal and the recipient was often as big as a foot and a half. Any blood flowing through the tubes would cool quickly, and platelets that stuck to the sides of the device would obstruct the flow just as quickly as well. Moreover, the severity of a blood incompatibility reaction depends largely on the amount of incompatible blood transfused; the human body can handle a small bit of blood even as foreign as animal blood with just mild symptoms. If only a small amount of animal blood entered Coga's system, this would go a long way toward explaining why Coga and Denis' patients remained healthy after a procedure that could otherwise potentially have had serious complications and might even have been deadly. A mild reaction would also help to explain the brief fevers that Coga experienced following the earlier trial and, again, after this latest one. Royal Society members had, however, their own explanation for the fevers, which were "justly imputed to his disordering himself by intemperate drinking of wine."[15]

The Royal Society had already made plans to perform a third experiment on the "extravagant" Arthur Coga. Lower's *De corde,* a treatise on blood transfusion that the surgeon published several years later, confirmed this: "In order to make further experiments on him with some profit also to himself, I had decided to repeat the treatment several times in an effort to improve his mental condition." But to the Royal Society's great surprise and disappointment, Coga refused under protest that he had been transformed into a sheep. Writing under the name Coga the Sheep (*Agnus Coga*), he complained that the virtuosi had "transform'd him into another species" and left him penniless. Since their sheep's blood had caused "the loss of his own wool," he explained that he would submit to future experiments only if the Royal Society would transform him entirely into a sheep ("without as well as within"). The letter was signed: "The meanest of your flock."[16]

When the heavy-drinking Coga claimed he been transformed into a sheep following his transfusions, he likely did it at the dictation of adversaries of the Royal Society, which was frequently mocked for its excessive enthusiasm for unconventional experiments.[17] "The coffee-houses," according to contemporary writer John Skippon, had "endeavored to debauch the fellow, and so consequently discredit the Royal Society and make the experiment ridiculous."[18]

The playwright Thomas Shadwell similarly took comic aim at the transfusion experiments performed by the Royal Society in his satire *The Virtuoso*. The play's main character, Sir Nicholas Gimcrack, tries their most famous experiments with comically disastrous results. Quoting portions of the transfusion reports published in the *Philosophical Transactions* nearly verbatim, Gimcrack replicates the Coga experiment in his own home. The results go beyond any of those described by Coga in his drunken revelries: Gimcrack "tranfus'd into a human vein 64 ounces of sheep's blood." And his patient became "fully Ovine, or Sheepish; he bleated perpetually and chewed the Cud. He had wool growing on him in great quantities, and Northhamptonshire Sheep's tail did soon arise from his anus, or Human fundament." The charlatan doctor resolves to make "a Flock" of human sheep. "I'll make all of my clothes from 'em," he exclaimed, " 'tis finer than a Beaver."[19]

Men like Boyle, Lower, King, and Clarke pursued transfusion with focused seriousness, as part of an intellectual and scientific puzzle that demanded answers. Yet for others transfusion was still another comic example of the ways in which natural philosophers had pushed experiments beyond the realm of the practical and into that of the ridiculous. Still, there seems to have been little outrage, at least in England, that transfusion was something morally corrupt or frighteningly wrong.

In France, of course, the reaction to such experiments could not have been more different. Transfusion, and more particularly Denis' experiments, had sparked angry reactions in the Academy of Sciences and the Paris Faculty of Medicine. Denis' experiments were nothing short of heretical, given the deep intersections between alchemy, transmutation, and Protestant approaches to medicine. At the very least Denis' imitation of the English was treasonous. An untalented imitator of the Royal Society, his detractors claimed, Denis had proved his allegiance to English science, rather than that of his fellow countrymen. And as Claude Perrault proclaimed, the transfusionist was clearly "prejudiced by the authority of the foreigners who had approved of transfusion."[20] But for Denis the publicity his experiments brought was much more important than originality, the advancement of science, or national allegiance.

Both Denis and Montmor were eager to continue building their reputation in Paris as contrarian outsiders to the king's nascent Academy of Sciences and the traditionalist Faculty of Medicine. To do so Denis stayed happily in copycat mode. The English had transfused the mentally ill Coga with sheep's blood. And now Denis began preparations to transfuse the legendary madman of Paris: Antoine Mauroy.

While we have few details about how Mauroy eventually came under Denis' care, we do know it happened shortly after the transfusionist, his faithful surgeon, Emmerez, and Montmor met formally to discuss next steps for Denis' transfusion experiments. Sometime in late November 1667 the men schemed to pluck Mauroy from the streets—by force, if necessary—and prepare him for his cure. Under the guise of taking pity on him, Montmor instructed members of his guard to find the madman and bring him to his estate.

Though Mauroy was well known in the Marais district, it had not been easy to track him down. The narrow city streets teemed with life and the stench that went with it. The streets collected the city's trash. They also collected its rejects: the penniless, the sick, and the mentally ill. Antoine was only one man in the capital city's sea of unfortunate souls.

Mauroy's insanity had been born of deep disappointment, the kind from which few gentle souls recover: He had formerly been a valet to the illustrious Marquise de Sévigné, among the most elite of the who's-who of literary Paris and a regular presence in the young King Louis XIV's court. Her comments, even in passing, could catapult an author into the stratosphere of high society or condemn him to anonymity. The once-affable Mauroy enjoyed a bird's-eye view of the lovely ladies who frequented Sévigné's salon. A man of humble origins, he should have restricted his pleasures to the earthier young women in the bustling kitchens or to those who carried the washing down to the river every day. But somewhere behind the pleasant smiles he shared with the more beautiful visitors, the then-twenty-four-year-old valet hoped to be able to romance his way into a considerable fortune. Yet he misjudged the strict rules of social mobility—or rather, social immobility—that dominated France in the late seventeenth century. Mauroy's failed romance was a heartbreaking reminder of his low standing; his beloved, whose name is lost to history, married a man of more suitable birth. To his dismay he would never be anything more than a lovelorn valet, the laughingstock of the Marais.

His frenzy came on quickly and brutally. Tearing his tailored uniform into pieces, the wretched Mauroy ran naked and screaming into the streets, threatening and swearing at passersby.

Madame de Sévigné herself called in doctors to cure the man. Barber-surgeons performed numerous bleedings to empty the noxious humors in the blood that had unsettled his mind. He was given cooling compresses and calming foods. Nothing seemed to work. Mauroy's behavior turned violent. He began setting fire to homes and threatening to kill the noblemen whom he blamed for his downfall. Sévigné had no choice but to ban Mauroy from her home for good.

Such exile proved too much for the fragile Mauroy to bear. Not long after, he was found in front of the gallows of the old Marais temple with a cord around his neck, howling that he would hang himself. A neighborhood nuisance, Mauroy inspired compassion in the upper classes. One lady of the court, Madame Commartin, had taken pity on him and ordered her valets to bring Mauroy to her home. She called in physician after physician, surgeon after surgeon, in the hope of curing the poor man. One physician had prescribed a series of bleedings from the feet, the arms, and even the head: eighteen in all.[21] Constant bathing was also ordered. In healthy persons, bathing had long been avoided because it was believed that water could seep into the body and weaken otherwise strong humors. Water was reserved for only faces and hands, while perfumes were rubbed vigorously over the rest of the body both to mask odors and to purify the disease-causing corrupt air with which the person came in contact.[22] In cases of madness, however, cold baths were thought to shock the patient into a more suitable mental state and to cool the vapors that were troubling the mind. Forty baths later, it was more than clear that traditional methods to cure Mauroy had failed. It was not long until the man could be seen running, naked once again, through the streets.

As the guards roamed the quarter in search of Mauroy, Montmor worked his few remaining connections to secure a temporary place for the former valet in a nearby hospital. Right across the river from the Marais stood the largest hospital in Paris, the Hôtel-Dieu de Paris; it was also the oldest. Literally "God's home in Paris," the hospital still sits where it did then, just adjacent to Notre Dame. The Hôtel-Dieu housed the swelling ranks of the city's sick, its indigent, and its insane. By boarding Mauroy in a nearby hospital, Montmor would have removed him from the streets—but he would also have condemned him to an overcrowded cell and to the constant drama of disease and death that defined all early European hospitals. Only those select few with ample resources could count on a bed—which was always shared by at least one other patient. But as it was, the hospitals were full, and Montmor's waning influence could not open a place for Mauroy. Instead the nobleman arranged lodging for the mentally disturbed man in a hostel on the nearby rue de Beaubourg, where he was locked in a small room and fattened up until the moment was right for his transfusion. And as Denis prepared for what would be his final transfusion, the transfusionist's most dangerous enemies were plotting his downfall.

Chapter 13

MONSTERS AND MARVELS

If transfusion was intended as a cure for insanity, it could also take sane men to the very brink of violent madness. Henri-Martin de la Martinière was one of those men. A self-trained doctor, Martinière was a man of deep faith and great passion; he was also a man of impressive hyperbole and rich imagination. Denis' detractors in the Academy of Science and the Paris Faculty of Medicine considered the transfusionist to be a dangerous renegade. Soon, the same—and more—would be said about the man who would become Denis' greatest foe.

If there was ever a man who would have seemed likely to support Jean-Baptiste Denis, it was Martinière. Born just a year before Denis, in 1634, the thirty-three-year-old Martinière was similarly restless by nature. At the age of nine he had run away from his home in Rouen and soon found a place for himself in a military encampment near Geneva. The boy could usually be found tagging behind the resident barber-surgeon in the camp's makeshift infirmary. There, Martinière learned how to keep patients bandaged and well bled, pull rotting teeth, and when occasions presented themselves, remove bullets and amputate battle-wracked limbs. His life among the French soldiers came to a sudden halt three years later when, at the age of twelve, he and his regiment were captured by Spanish forces. After much fast talking Martinière negotiated his way onto a ship bound for the East.[1]

Martinière had been on the Portuguese ship for just two days when the captain spotted a fleet of ships in the distance. In this no-man's-land along the Barbary Coast, corsairs ruled. The captain called his crew to the decks and passed around bottles of hard alcohol. Each man, including the barely teenage Martinière, took a swig. They would fight the pirates off with every ounce of strength they had in them—and to the death. Fifteen minutes later they were surrounded. Martinière's shipmates set off every cannon they had and, within moments, the air was black with soot. One of the corsair ships splintered in two and groaned as it sank to the bottom of the sea. Fearless, the young Martinière crouched behind the balustrade of the ship and deftly added to the gunshots volleyed in every direction. But outnumbered and insufficiently armed, his ship began its slow descent into the blue-gray water.

The pirates brought the healthiest-looking men aboard, planning to make a tidy profit selling the foreigners as slaves. But the corsairs had lost their surgeon, and their captain was looking for a replacement among the captives. The lanky Martinière stepped forward, declaring confidently that he was the surgeon of the captured ship. The pirates bellowed with laughter. The captain growled at the boy and asked if his master, the ship's surgeon, had been killed. Martinière growled back dismissively, "No, because I'm still alive."[2] The captain looked the boy up and down one more time and nodded to one of his men, who tossed the surgeon's tool kit to Martinière.

Martinière's travels afforded him a view of the worst of what humanity had to offer. During a stop in Egypt he had witnessed the dark side of medicine while stocking up on supplies at a local apothecary shop. There, in a back room, he saw piles of desiccated human bodies layered one on top of another. In a corner a man was removing the brains and internal organs of a fresh body. The body was then filled with a black sticky liquid, wrapped in dressings, and left to dry. The "mummies" were destined for export to Europe, where they were prized for their purported curative properties by even the most highly respected physicians. Small bits of dried mummy flesh, ingested either whole or powdered, were believed to cure a wide range of ailments, such as headaches, paralysis, epilepsy, vertigo, earaches, sore throats, scorpion stings, and incontinence. [3] Like most Europeans, Martinière had always believed that the mummies had been pulled from ancient sands. Instead the apothecary-turned-body-trafficker admitted that smallpox, leprosy, or plague provided a steadier supply of bodies.[4] And when that was not sufficient to meet the gluttonous demands of the West, there were no doubt other, more criminal ways, of finding "volunteers."

As a young boy Martinière spent his nights paralyzed by fear as he overheard his pirate captors tell stories of grayish red sirens with hair as thick as horse manes and wings on their backs, and of two-headed Hydras whose poisonous bite rotted human flesh.[5] Men could be just as terrifying as monsters, it seemed. And, for the pirate-surgeon, monsters would forever lurk around every dark corner.

At the age of sixteen Martinière was rescued from the corsairs by Maltese soldiers. He continued his nomadic life much as he had before, connecting with others who could provide food, lodging, and protection. From Malta he traveled to Rome and offered his services at a hospital there. From Rome he made his way slowly back to Rouen, where he attended medical school and became officially what he had already been for years: a physician. Records of his life in Rouen and in the years preceding his involvement in blood transfusion controversies are scant. However, there is little doubt that Martinière's ability to work his connections in whatever community he happened to find himself certainly proved useful.

Barely five years after earning his medical degree, Martinière had negotiated an appointment as a physician in the court of Louis XIV, which offered him the same status and privileges as others who had trained in the much more elite—and certainly more conventional—medical faculty at the University of Paris. Yet, like Denis, Martinière floated on the periphery of the Paris scientific elite. While his talent for insinuating himself into any situation had come in handy, it had also come up short. He had been given a rare exception to be able to practice medicine in the capital, but without a diploma from the University of Paris, he would never teach there. And as much as he may have wished to count himself a full member of the Paris medical corps, the opportunity to interact with those in its inner sanctum would present itself only too infrequently. Still Martinière, now a credentialed and practicing Paris physician, would never be able to shake the education he had received among the pirates. He was always looking over his shoulder for potential danger—and blood transfusion loomed large as one of the greatest threats to mankind he had ever seen.

Martinière knew there could be no proposal more macabre than blood transfusion. He worried about where physicians planned to get their blood sources, if they should ever decide to transfuse humans to humans. Would physicians—like the Egyptian apothecaries who trafficked "mummies"—become little more than "buyers and sellers of human blood," as they brokered deals for rich men who wished to buy the blood of beggars? Martinière shuddered as he imagined in graphic detail a host of violent scenarios. Children would be whipped so that their blood vessels would warm and fill with as much blood as possible—then they would be offered to men desperate to cure horrific illnesses such as plague and syphilis.[6]

Blood transfusion, he believed, was just a first step to the worst imaginable crimes against humanity—and certain damnation. "Men will cut one another's throat to preserve their life," wrote Martinière.[7] Soon, there would be nothing keeping people from "bathing several times over" in the blood of innocent victims. Or worse: cannibalism. "Whoever is unscrupulous enough to fill their veins with the blood of another," he declared with an almost palpable shudder, "will find little trouble in eating human flesh to heal himself."[8]

On a chilly fall night in 1667 Martinière had a dream about transfusion that would set him on a divine mission to right the wrongs men like Denis seemed bent on committing. As he would later claim in his many spirited and rambling writings against transfusion, a beautiful woman with bright eyes came to him while he slept fitfully that night and stood in front of his bed. She whispered to him reassuringly, and delicately rinsed his face and eyes with clear, clean water. From the corner of his eye Martinière also saw a young man step toward him holding a lyre in one hand and an archery bow in the other. "Who are you?" gasped Martinière. "I am Apollo," the young man replied. "Son of Jupiter and father of the great physician Asclepius. The goddess Truth has washed your eyes so that you can see what I want to show you. Engrave this into your mind so that you can spread the message to the rest of the human race."[9]

At Apollo's command the goddess whipped back the bed-curtains. Martinière gasped in horror. His bed now dangled from a steep precipice. A hideous smell of death emanated from the deep pit below. Leaning from his bed, he squinted as he struggled to make out the scene. A group of natural philosophers were concocting experiments. Their purpose: to ensure their fame and to get closer to finding the secret of immortality. In one of these experiments a natural philosopher grabbed a helpless animal and deftly sliced its tail off. Taking a syringe, he injected the animal's tail with so much milk that white liquid began to spurt from its eyes, nose, and ears.

The philosopher turned to his colleagues and declared proudly, "You are witness to this new miracle that I have invented. Those who can no longer eat can now introduce food into their veins." The others in the group nodded approvingly. Then, like ravenous beasts, the philosophers each grabbed pairs of animals and began a frenzied rush of transfusions. Lion's blood flowed into lamb veins; lamb's blood into wolf veins. Martinière's dream replayed the history of early blood experiments—from Wren's infusions of milk, beer, and blood to Lower's hopes of intravenous feeding. But it was the final act, evoking as it did Boyle's hopes for interspecies transmutations, that brought the greatest of horrors. Without warning the macabre school of philosophers turned their bloody lancets on one of their own. Their human victim was bled dry, and his veins filled with cow's blood. Martinière watched as the philosopher transformed, slowly but surely, into a large cow. The philosophers bemoaned in horror the plight of their colleague, as they worked desperately to undo the spell that their transfusion had cast on him. Then the curtains closed; the room went dark.[10]

As the sleeping Martinière battled monsters created by these philosopher-transfusionists, his footman gently opened the chamber door. He held in his hands copies of three letters: Denis' letter to Montmor, which announced his successful transfusions;

a second letter by the influential Paris doctor Guillaume Lamy to the venerable physician René Moreau, which denounced the procedure vehemently; and a third by a Monsieur Gadroys, which took exception to Lamy's criticisms and defended Denis.[11] The letters had been sent by an unnamed colleague who was asking for Martinière's opinion on the matter.

Martinière's nightmare soon became reality as his eyes scanned the pages. In the first letter Jean-Baptiste Denis crowed about his successes transfusing dogs to dogs, then animals of different species, and finally the young boy and the butcher. It was a copy of the same letter Oldenburg had received, and it provoked in Martinière similar outrage. His concerns about interspecies blood transfusions were confirmed, and heightened, as he read the second letter.

Guillaume Lamy, a well-respected and influential member of the Paris medical faculty, left no room for debate when it came to his thoughts on blood transfusion. The procedure, he wrote, was more than just "a new way to torment sick people."[12] To transfuse animal blood into human veins, he argued, would have "very grave effects." The flesh and blood of every creature, humans included, contained different "particles" that produced the different qualities and characteristics that could then be used to define the creature's distinction from the rest. The blood of some animals contained particles useful to grow horns, so they grew horns. Cows had particles in their blood that made them stupid, Lamy explained, so they were stupid.[13]

Both in the letter and throughout his writings over the course of his career, Lamy dismissed the idea of finalism, which held that both animals and humans were formed as they were and behaved as they did because of some divine unalienable plan. If man turned out the way he did, it was because God had offered a menu of qualities and characteristics possible in all of his creatures, but it was only—literally, for Lamy—by the luck of the draw during reproduction that humans turned out as they did. "As three dice," Lamy explained, "thrown on a table form of necessity some of the numbers that are between three and eighteen, without possibly forming either more or less, so in the same way the particles of the seed [*semence*] ineluctably make some man, without being able to produce a body of another species."[14] It was only once the qualities of a given species had been set that the creature found a way to make use of them. As Lamy emphasized, eyes were not made for seeing; rather, beings see because they have eyes.[15]

Lamy's opposition to blood transfusion sprang from the fact that it mixed particles that would now fully "belong" to separate species following this complicated game of luck. He warned in his letter that mixing the blood of a cow and a human would lead to "pernicious effects." It risked infusing humans with particles that would give them, like cows, "heavy and slow minds." A bitter critic of Descartes and his theories of mind-body dualism, Lamy also feared deeply that the human soul was also at risk. Lamy was a materialist who believed, much like Thomas Willis, that the soul resided—in both animals and humans—firmly in the body itself. Mixing the blood of species meant not just transferring physical appearance and behaviorial inclinations from donor to recipient, it also risked transferring the very qualities of the soul from one to the other.

Lamy's letter had been strategic. Before being distributed broadly, it had first been delivered in manuscript to René Moreau, a well-known member of the Paris Faculty of Medicine whose loathing for Montpellier doctors was legendary. Nearly thirty-five years earlier the now-elderly Moreau had been one of the chief architects of an intellectual assault against Théophraste Renaudot, an outspoken advocate of Paracelsan chemical medicine. A graduate of the University of Montpellier, Renaudot began practicing in the capital city without the permission of the University of Paris medical faculty. He created a *bureau d'adresse*, a central "address" or agency, where the homeless could request financial, legal, and especially medical help. The need was great, and Renaudot's services were in high demand. His medical center regularly dispensed doses of antimony and other chemical remedies of which the medical faculty disapproved, and as it grew, it threatened to overtake the influence of the Paris Faculty of Medicine on the hearts and minds of the populace. Unwilling to put up with this threat to their dominance, in 1640 the faculty filed charges against Renaudot for practicing medicine without their approval.

In 1641 Moreau had written a spirited *Defense of the Faculty of Medicine of Paris.* In it he argued passionately that Renaudot's many diplomas were not worth the paper they were printed on and conveyed no right to practice in Paris. Moreover he had been a mere child when he received his medical training at the age of nineteen and clearly did not have a firm-enough grasp on medicine to actually practice on living, breathing patients. And finally Moreau reviewed nearly every argument imaginable against antimony, showing that even Renaudot's own colleagues from the South of France were not convinced that it worked.[16] After several years of legal wrangling Renaudot was forced to abandon his practice and was later officially stripped of all privileges. Not long after, the emboldened Paris Faculty of Medicine issued a formal condemnation of *all* Montpellier doctors practicing in Paris.[17]

The Renaudot trial was a euphoric triumph for Paris physicians—and one that Lamy hoped could be repeated now, in the case of Jean-Baptiste Denis. By sending his letter to Moreau, Lamy signaled to the faculty at the University of Paris that it was now time to go on the offensive against this most recent Montpellier-trained menace.

Martinière was convinced that the timing of the letters, following as it did his horrific dream, was not a simple coincidence: It was a divine call to action. "To allow foreign blood to enter one's veins," Martinière resolved, "is to bring about a bloodbath, a most inhuman remedy," one that "will attract the ire of God."[18] For Martinière, a physician in the court of Louis XIV, the boundaries between nightmare and reality were overlapping in the most terrifying of ways. He had kept abreast of English questions about transfusion's potential to transmute species. And circulating as he now did among the Paris physicians, he was well aware of the anger that Denis' experiments had caused.

But for Martinière the dangers of transfusion were not simply abstract conjecture. The idea that humans and beasts could be merged in novel ways seemed frighteningly real. As Martinière battled monsters in his dreams, he was also reliving his greatest childhood fears. And if his brutal education on the pirate ship had taught him anything, it was that when one had to choose between fight or flight: Fight.

Chapter 14

THE WIDOW

While Mauroy was sequestered and awaiting the transfusionist's knife, his wife was roaming Paris in search of him. A villager unaccustomed to the bustle of the big city, Perrine wove her way awkwardly through a crush of bodies and carriages. The noise was deafening: Hawkers clanged bells announcing that they had brandy for the men trudging to work, town criers shouted the day's news, quarreling neighbors screeched at the tops of their lungs.[1]

Perrine knew the cold fact behind her husband's violent outbursts. She had come to understand it long ago: He did not love her; he never had. He had longed desperately to marry a noblewoman whose love could never be requited, and Perrine paid daily for his disappointment. His first fit of "extravagance" set him on a rampage that lasted ten months. As he began to come slowly to his senses, Mauroy discovered that his marriage had been arranged to a woman in his village, ten miles from Paris. Mauroy's family had been convinced that married life would provide the stability he needed; they had even been able to persuade the bride that Mauroy's madness had been due to a temporary illness. Young and trusting, Perrine believed them. But within the first year of his marriage, Mauroy's anger and disappointment again drove him to madness. No longer naive, Perrine realized that she had been duped and was now consigned to a life of regular beatings.

No matter how many doctors or clergymen came to the Mauroys' home to cure him, his outbursts were unremitting. Mauroy was not in a position to support his family, and whatever was left of his wife's dowry had been spent long ago at the apothecary and on doctors' fees. Perrine begged a few fellow villagers to help her restrain her husband in the home they shared. After a mighty struggle Mauroy found himself tied tightly to the bed. Perrine did not do this to protect her husband from himself; she did it for her own safety. In the worst days of his angry fits, she feared for her life. But despite his wife's best efforts, Mauroy found a way to slip out of his restraints in the late fall of 1667. He ran straight to Paris, where he was soon intercepted by Montmor's guards and held alone in a small room for his history-making transfusion.

Mauroy found himself tied up again—this time strapped to a chair in Montmor's large meeting hall—and surrounded by well-dressed strangers. While Montmor watched euphorically, Denis and Emmerez transfused Mauroy with about six ounces of lamb's blood. The second transfusion a few days later proved to be more intense for the patient. He felt the same heat in his arm, his pulse raced, and his face quickly became covered in sweat. The hours immediately following the procedure had also been perilous. Mauroy experienced a bout of debilitating fever, nausea, diarrhea, nosebleeds, and urine that was as black as "chimney soot." But again, just days after the experiment, his body seemed to recover. And now his sane behavior served as proof of transfusion's benefits.

News spread quickly throughout Paris. The crowded streets of the Marais buzzed with excitement with the news of Denis' "cure." Given her husband's past, Perrine suspected that he had fled to Paris, where she wandered the streets in search of him. If it is not easy to understand why Mauroy's wife would have searched for him, given how much she feared him, it would also not have been easy to be a woman of little means on her own in late-seventeenth-century Europe. Any husband, even one as uncaring and unbalanced as Mauroy, was perhaps better than no husband at all.

As Perrine heard gossips tell stories about the madman now in Montmor's care, she knew without a doubt that they could only be talking about her husband. So, on Christmas Eve, just four days after the second experiment, the penniless Perrine Mauroy passed through the gates of Montmor's comfortable residence. She marveled at the richly appointed home and soon marveled as well at the dramatic change in her husband, who was now "of a very calm spirit." Visitors who had known Mauroy during his employment by Madame de Sévigné also flowed into Montmor's home. Each in turned confirmed that the man "was restored to the same state he used to be in before his Phrenzy."[2]

Denis set to work on broadly publicizing his successes. As the holiday season came to a close, the transfusionist sent letters to every corner of Europe. First on his list was Henry Oldenburg. While the secretary of the Royal Society likely continued harboring ill will toward the French transfusionist, the society's successes with Arthur Coga had done much to temper animosities. To Denis' great delight, his letter describing Mauroy's two successful transfusions was published, in full and in translation, in the February 10, 1668, edition of the *Philosophical Transactions.* The French original was printed shortly afterward in the *Journal des sçavans* as well.

No one could have been more delighted than Montmor. His

reward came in the form of correspondence between Denis and Samuel Sorbière, the secretary of Montmor's now-defunct academy. It was Sorbière who, four years earlier, had initiated the move to dismantle Montmor's private gatherings in order to replace them with the king's academy—and Montmor still bristled at the thought. But Sorbière had experienced his own setbacks. Jean-Baptiste Colbert, Louis XIV's prime minister, recognized that Sorbière was nothing more than an opportunist. Having not been invited to join the king's new Academy of Sciences, the dejected Sorbière was now in Italy, grasping at straws in an attempt to woo scientific power brokers there. Transfusion was quickly gaining ground in Rome and Bologna. The ultimate goal for Italian natural philosophers was, as it had been for the French and English, to perform their experiments on humans. But they still had a long way to go. Seeing an opportunity, Sorbière was now begging Denis for details of the activities that had taken place in Montmor's home.[3] Denis obliged, bragging with even more bravado than usual about his successes at the Montmor estate. This delighted Montmor immeasurably. Revenge was sweet. Sorbière had publicly humiliated him in front of his academy. Yet now it was he, not Sorbière, who was the talk of scientific Paris. Clearly his bet on Denis and his efforts to restore glory to his private academy seemed to have paid off.

Denis visited his patient every day in the week that followed the first two transfusions. When the physician declared with great pride that Mauroy had been fully cured, Mauroy and his wife eventually returned to their modest home. Perrine was hopeful that the past was behind them but had a lingering sense of dread that her husband's newfound calmness was temporary. She was right. Mauroy remained in "that good condition" for about two months. The man's state of health and mind changed abruptly when—Denis explained, eager to cast blame somewhere—the too frequent company with his wife and his debauches in wine, tobacco, and "strong waters" (alcohol) had cast him into a very violent and dangerous fever."[4] With his brain warmed by the fever's vapors, the mad man's ravings soon returned, even worse than before.

Early one morning in mid-February, Madame Mauroy left her village home and rushed to Paris to urge the transfusionist to perform a third surgery. Something in Perrine's demeanor had completely changed. The once-timid woman showed up on Denis' doorstep in a fury. She was filled with a boldness that made her nearly unrecognizable to those who knew her only as Mauroy's battered and mousy wife. The source of her newfound confidence was still a mystery, and it would not be unveiled until many months later. But clearly she felt empowered. She snapped at his housekeepers in a voice filled with a strange and urgent authority: Denis would meet with her—or else. With a belligerence that surpassed her social standing, she threatened to have a Paris judge force Denis to transfuse her husband if he would not do it of his own accord.

Informed of the unannounced visitor, Denis spoke with Emmerez, and the men decided to make a house call to the Mauroy family home. Denis' carriage slogged through the muddy paths that led to Perrine's ramshackle abode in the middle of what seemed like nowhere. Perrine walked briskly up to the men and, sparing few words, ushered them into the one-room house. Mauroy howled at Denis and Emmerez as they entered; he had been tied tightly by his arms and legs to the couple's bed. In the center of the room surgical tools and bloodletting bowls neatly lined the wobbly dining table. Perrine gestured outside to a calf that was tied to a nearby fence, and she ordered them to begin the procedure immediately. Nothing about the situation was right; Denis could sense it. Perrine certainly did not have the resources to buy such tools, nor could she have known what specific implements they required for the procedure. And the calf? The Mauroys had barely enough money to feed themselves, much less to buy an animal specifically for the procedure. Moving slowly back toward the door, Denis and Emmerez informed the woman firmly but politely that her husband was in no condition for the operation.

Before the men had a chance to leave, Perrine moved toward the light of a nearby lamp. Turning to the men, she displayed her mottled face, black and blue from her husband's beatings. Then she crumpled to the floor in tears. She pleaded with them not to leave without giving her "the satisfaction of having tried all possible means to recover [her] husband."[5] Her distress was artful, Denis explained later, and the men agreed against their better judgment to transfuse Mauroy one more time. Reluctantly Emmerez first passed a narrow tube into the vein of Mauroy's arm and capped it off. Next he opened a vein in the man's foot because, as was now commonly believed, old blood needed to be removed when new blood was infused. Suddenly Mauroy's body shook in a "violent fit." Every limb of his body trembled wildly. "There issued no blood out of the foot, nor the arm," Denis later insisted emphatically. And Emmerez had no choice, according to Denis, but to remove the tube from the man's arm and stitch him back up without opening the artery of the calf. When the men left Perrine's house they were troubled by what they had seen. Mauroy died the next day.

As soon as Denis heard the news of his patient's death, he summoned Emmerez to his side. The companions replayed every detail of the last transfusion. Mauroy was clearly not in his right mind when they arrived at the Mauroys'. But other than the unexpected seizure, he had shown no other signs of illness or infirmity. There had to be a reason for his death; it was surely not the result of a transfusion. Of this Denis could not have been more certain: No blood had actually been transfused into Mauroy. The calf's artery had not even been opened when Mauroy began having seizures. And the two men, Denis insisted, had left without transfusing Mauroy. Yet, the man was dead. It did not make sense.

In a quest to discover the truth, Denis and Emmerez climbed into a carriage next morning to visit Mauroy's widow. They wanted to know what her husband's last hours had been like, what his symptoms had been; after this they would perform an autopsy. When Denis and Emmerez came face-to-face with the man's widow, it was clear that she was not at all pleased to see them. She was now as difficult to reason with as her husband had been. Denis tried to coax information, any information, from the widow about her husband's final hours. But Perrine was hostile and unhelpful. Mauroy's thin body, blue and cold, was stretched out on the couple's rope bed. Denis visually inspected what he could of the corpse, as Emmerez placed his tool kit on the table in the center of the room. As soon as Perrine understood that the two men intended to perform an autopsy, she exploded. Enraged, she chased them from her home. Unwilling to back down, Denis yelled over his shoulder at Perrine as he stomped to the carriage. He made it clear that they would return the next morning "to do the thing by force."

As anxious and now angry as Denis was, he likely sensed that one additional day would not make a difference. He hoped that the widow would find a way to calm herself by morning and allow them to get to the bottom of the mystery in a more rational way. Plus, time seemed to be on their side. Denis was certain that the penniless widow could ill afford the burial expenses; Antoine Mauroy would still be there in the morning.

Denis underestimated Perrine Mauroy. As soon as the transfusionist's carriage was out of sight, she sprang into action and began making frantic arrangements to bury her husband with "all speed." What Denis failed to remember was that the widow Mauroy had been surprisingly able to cull the resources she needed when she put her mind to it. She had somehow found a way to procure a calf and the necessary surgical instruments for him to transfuse her husband for the third and last time. And once again, by some mystery, she was able to buy a coffin for her dead husband and pay the gravediggers. Denis and Emmerez returned the next morning as promised, but Mauroy's corpse was gone: It was already in the ground.

In the days that followed the widow's odd behavior, the transfusionist reviewed every conversation he had had with Perrine and her husband while Mauroy was still alive. There was no doubt that the widow was hiding something. At various moments before and after the first transfusions, Mauroy had howled in panic—convinced that his wife was plotting his death. At the time he and Emmerez had brushed his fears aside as the wild ravings of an insane man. Now they wondered if it was possible that underneath the exterior of this seemingly fragile woman, as bruised and as battered as she was, there lurked the heart of a murderer. But without a body, Denis worried that there would be no way to discover the truth.

Mauroy's death offered Denis' detractors a most welcome chance to take the arrogant transfusionist down a notch. The number of transfusion's foes was growing daily, especially now that the stalwart members of Paris Faculty of Medicine, Perrault and Lamy, had turned their resistance to the procedure into a cause célèbre. Then there was the pirate-turned-Paris-physician Martinière, whose invectives against transfusion were becoming shriller with each passing day. In the weeks following his vivid dream, Martinière had published at his own expense a series of treatises and letters to prominent doctors, court members, and parliamentarians decrying transfusion. While not a member of the Paris Faculty of Medicine, Martinière had found ways to make a name for himself among its members. It was not hard. Martinière's animated and hyperbolic stance toward transfusion had been coupled with the manners of a man who had spent his youth among unruly and uncouth seafarers. The Paris physicians could hardly disagree with what he was saying. And Martinière had recently discovered a kindred spirit in the influential Guillaume Lamy, who was leading the Paris Faculty of Medicine's charge against transfusion.

As news of Mauroy's death spread, these and other *anti-transfuseurs* reported the news gleefully in their correspondence. In a chorus of "I told you sos" they accused Denis of being too quick to take credit for his success. Accusations against the transfusionist were now also swirling openly among the highest-ranking members of the king's court. On February 15, 1668, Louis XIV's secretary, Henri Justel, was finishing a lengthy letter to Oldenburg with a summary of the latest French gossip. He had just reached for the pounce powder to dry the ink when he was told of Mauroy's death. He was sure that Oldenburg would want to know right away. The courier was waiting to speed the letter across the Channel, so Justel reached for a small slip of paper and appended the news to his letter. "After my letter was written," Justel wrote briskly, "I learned that the madman who had received the blood transfusion is dead. So Monsieur Denis boasted inopportunely of his cure. He should have waited."[6] Ten days later Justel further reported to Oldenburg that transfusion was "being cried down."[7] "Their mischance," Justel explained smugly, "will discredit transfusion and no one will dare to try it in the future on men."[8]

Chapter 15

THE AFFAIR OF THE POISONS

Denis had always delighted in stirring up controversy, yet the transfusionist knew better than to inflame his detractors, who were hard at work concocting a case against him. In the months that followed Mauroy's death Denis had little choice but to remain silent. Without a corpse, he knew that he could not offer a definitive explanation for the man's demise. Still, Denis stood firm in his conviction that transfusion had not killed Mauroy. In fact he became more convinced than ever that his patient had been the victim of foul play.

Hope came two months later, in the form of Perrine Mauroy. Standing on the doorstep of Denis' home on the Left Bank, Perrine looked up at the transfusionist with contrition. She claimed that several men, physicians all, had circled her home in the days following her husband's death. They waved large amounts of money and "did extremely solicit her to bear false witness against Denis"; they had even spoken to all her neighbors so that they might persuade her to bear false witness in the courts. The repentant Perrine told Denis that she had refused, "knowing her obligations to [Denis] for having relieved her husband freely." Denis expressed his own gratitude to Perrine for her excellent judgment. Still, he suspected that he had not yet heard everything that she had come to say.[1] He was right.

Perrine described with much pathos her plight as a penniless widow. Turning on as much charm as she could, she told Denis that she had done everything she could to resist the money that they had offered her. Perhaps he might be able to help her? She had been inside Montmor's sprawling compound and knew that Denis benefited from the generosity of his patron. Certainly he and *Montmor le Riche* would see fit to help her out so that she would not have to accept the bribes of others. Unmoved, Denis snapped at Perrine. He told her angrily that "those Physicians and herself stood more in need of the transfusion than ever her husband had done," and that he did not care for threats.[2] Perrine left in a huff.

The widow's story of outside involvement, by other physicians no less, made it clear to Denis that a plot was brewing against him. Something very similar had happened to Nicolas Fouquet a few years earlier. The superintendent of finances had dared to put on an unforgettable display of wealth and hospitality during a visit by Louis XIV to his estate at Vaux-le-Vicomte. He was not punished immediately for his transgressions. Instead the king and his soon-to-be prime minister, Colbert, coordinated a surprise arrest a full month after the fateful party. Now Fouquet was wasting away in solitary confinement in the dungeons of Vincennes.

Denis knew his ambitions had led him directly into the path of the sun: Louis XIV, the Sun King himself. The transfusionist had gone directly against the will of the king's Academy of Sciences as well as the major players at the conservative Paris Faculty of Medicine. It is not clear whether Denis' next actions were intended to preempt a strike that he knew was coming or whether it was a last show of his characteristic arrogance. Perhaps it was both. What is certain, however, is that Denis still believed unequivocally in the potential of blood transfusion. He resolved that he would get to the bottom of things and restore not only his good name but also the procedure on which he had staked his reputation.

Not long after Perrine's unexpected visit to his home, Denis made a visit to the fortressed Grand Châtelet, which stood menacingly on the Right Bank of the city. From the fourteenth century to its demolition in the nineteenth century, the Grand Châtelet was home to the main criminal and civil courts for the city of Paris and its surrounding villages. The massive building cast an eerie shadow over the river below. And behind its six-foot-thick walls the Châtelet teemed with activity as hundreds of lawyers, judges, notaries, plaintiffs, and defendants filled its many courtrooms daily.

Prime Minister Colbert had just recently brought the courts and the police system together under the same roof by reserving a special set of rooms in the overcrowded Châtelet for Nicolas de la Reynie, the newly appointed police chief of Paris. In short order Reynie put his mark on Paris as he tried to free the city from its infamous reputation as the crime and murder capital of Europe. Before Reynie's appointment, few people had been safe from the random acts of violence that defined daily life in Paris—not even one of the highest-ranking judges at Châtelet, the "Criminal Lieutenant" himself. In August 1665 the Honorable Jacques Tardieu and his wife were murdered in broad daylight and in their own home, victims of an armed robbery. Colbert had appointed Reynie because he had seen enough, and he gave his new police chief the exceptionally broad powers needed to restore order to the violent city.

Reynie's presence in the castle-like compound only intensified

FIGURE 21: The notorious Grand Châtelet was both a major prison and the seat of the Paris courts in early France. It sat on the Right Bank of the Seine, directly across the river from the Conciergerie and Parliament. It was demolished in the early nineteenth century.

the Grand Châtelet's chilling reputation. Châtelet was legendary for its horrific conditions and deadly spectacles. The sprawling complex was infested with legions of rats that migrated from nearby slaughterhouses. The Châtelet also served as the city's main morgue, where the bodies of less fortunate prisoners were collected along with the many others that were regularly discovered in the streets of Paris or found floating in the Seine.[3] The Châtelet's stench was recognizable to every Parisian and hovered permanently in the air throughout the streets around the fortress.

In early Europe justice and death often went hand in hand. Many prisoners were crammed into windowless cells in the Grand Châtelet's largest turret. Others were left in *cachots* or *oubliettes* (from the French *cacher*, "to hide," and *oublier*, "to forget") in dungeons dug nearly five stories underground. Defendants in criminal cases who entered the Châtelet's courtrooms worried that they, too, could be sentenced to one of these prison cells—or perhaps lowered into the "pit." (The most notorious criminals were sent into what was likely a well-like shaft by means of a bucket and pulley. Unable to sit or lie down, they stood with water around their feet until they collapsed and died—usually after about two weeks.)[4]

Denis set out for the fortress intent on clearing his name of the accusations of ineptitude or, worse, murder that were now being made against him. In self-defense, he lodged a formal complaint against Mauroy's widow and her still unnamed accomplices. As his carriage approached the main entrance of the Châtelet, the transfusionist was confident that the police chief's commissioners and the judges there would be as appalled as he was by the widow's story of bribery and extortion. The coachman tried with difficulty to navigate the stream of bedraggled Parisians who elbowed one another for their place in line to meet with one of the police commissioners in order to report the transgressions of other city dwellers. When Denis stepped out of his carriage, the guard at the gate noted the physician's fine clothes and assured manner. In a matter of moments he was escorted to the Châtelet's main courtyard, up a set of massive stone stairs, and into a private room off the main corridor of the police headquarters.

In this room and away from prying eyes, commissioners met with "persons of quality" who either brought complaints against others, or who were questioned quietly about complaints brought against them.[5] Trying hard to keep his outrage in check, Denis told Commissioner Le Cerf his story. He described his medical experiments and explained with pride how he had transfused Mauroy successfully on two occasions, along with several others who were very much still alive to vouch for his skills. Denis crafted the case for his innocence and worked diligently to turn the commissioner's suspicions toward the bribing and conniving widow and those who had helped hasten the death of Antoine Mauroy. Surely, Denis argued, if the king was interested in keeping the streets clean of evildoers, the commissioner had little choice but to investigate the case. Finding sufficient grounds for concern, La Cerf forwarded the case to the Criminal Lieutenant, the Honorable Jacques Defita, for a full hearing.[6]

Courtrooms at the Châtelet were like theaters. A judge dressed in a flowing black gown and a perfectly coiffed wig sat high above the crowds on a narrow, oval-shaped proscenium. Lawyers for each side as well as a recording notary clustered around a table below and looked out at the men and women who had been brought in to tell their tales. On April 17, 1668, five people—Denis, Emmerez, Perrine Mauroy, and two of her neighbors—stood nervously in front of the table as they strained their necks to have a look at Judge Defita. Perrine and her neighbors seemed disheveled and out of place compared with the black-velvet-gowned lawyers who stood before them. Denis, on the other hand, wore an elaborately stitched waistcoat and expensive knee breeches and looked as if he belonged fully to the French court and its nobility. A balustrade behind them separated the space of testimony and judgment from a crowd of gawking spectators, agog with excitement and curiosity.

Judge Defita solemnly nodded to the two lawyers representing each side of the case. Sitting in the high-backed chair on the stage, he asked no questions. As was the tradition, he left the talking to his assistant, André Lefèvre d'Ormesson, who sat at his side. Barely twenty-three years old, Ormesson was still green. Like most young lawyers at the Châtelet, he came from an illustrious legal family and had been appointed to the post as a way to gain experience in the legal ranks before he took a position more worthy of his name. Denis could not have been more pleased that an Ormesson had been assigned to his case. André's legendary father, Olivier Lefèvre d'Ormesson, had risked his entire career to ensure a fair trial for Nicolas Fouquet. In the months following Fouquet's catastrophic party at Vaux-le-Vicomte, Fouquet's fate—and his very life—had been at the mercy of the courts. Serving as judge at the trial, the elder Ormesson refused to rubber-stamp the monarchy's case against Louis' former superintendent of finances. Colbert had expected the courts to move swiftly and deliver a death sentence. Instead Ormesson spent five long days in public deliberation and exposed the many irregularities he found in the case. Ormesson had the most serious charges dropped, effectively protecting Fouquet from the gallows. While Fouquet did receive a life sentence, Louis XIV and Colbert were nevertheless displeased. Not long after the ruling Ormesson "retired" from the court.

The younger Ormesson was well aware that the Denis case had many similarities to his father's famous trial. The Fouquet hearing had been a test of France's long-standing legal system in the wake of the new king's efforts to consolidate his power. Ormesson no doubt believed, like his father, that Denis—and transfusion more generally—deserved a fair trial. But it would have been foolhardy to overlook the political stakes of the case. Led by Guillaume Lamy and Claude Perrault, the University of

Paris medical school had made no secret of its outrage that Denis had been performing experiments without its express approval. The faculty had also made clear, through formal and informal routes, transfusion was to be stopped, quickly and for good. And if it took the death of a pitiful homeless man to make that happen, so be it.

When the members of the court had taken their places, Ormesson called his first witness. The room hushed as Denis approached the lawyer's table. He was asked to relate from memory the events leading up to Mauroy's death. Looking both Defita and Ormesson directly in the eye, the transfusionist described how he and Emmerez had transfused Mauroy on two occasions, each time with great success. In the two months that followed, he had carefully monitored the man's health and found him to be "in his good senses and in good health."[7] Denis described at length the difficulty that he and Emmerez had when they attempted the procedure for the third time. They had barely begun to bleed their patient in preparation for the procedure when the man began to have repetitive seizures, and as much as he had intended to perform the procedure, the man's condition prevented it. To no one's surprise Emmerez confirmed all of Denis' statements.

Ormesson then turned to the widow Mauroy. Existing records do not describe her demeanor, but we may imagine that she was terrified. A simple villager unfamiliar with the tight-knit social world of the Parisian upper classes, Perrine had undertaken an unthinkable odyssey. Just months earlier she had found herself in the Montmor estate which, despite the nobleman's change of fortune among scientists, remained one of the city's most opulent addresses. And now here she was in the capital's legendary home of justice—and death. Ormesson pushed the trembling widow to provide details about daily life with her husband in the short time between the first round of transfusions and his death. Perrine begged the lawyer to believe that she loved her husband. She had always taken great pains to respond to his every need. She fed him, she clothed him, and she prepared his eggs and broths.

Within weeks of the transfusions, Perrine testified, her husband's behavior became erratic. He went from one cabaret to another, where he drank, smoked, and flirted with other women. The dutiful Perrine claimed that she helped her husband nurse his hangovers. She mixed "strong water" (a form of alcohol) in his broth in order to help relieve his headaches. And, as final proof of her affections, she quietly told Ormesson that she had even bedded him four times in the weeks following his transfusions. Ormesson asked the widow if her decision to bed her husband was one of which his doctors would have approved. No, she explained, they had expressly forbidden the couple to have intercourse, for fear that the act would overheat the husband's blood and invite his frenzy to return.[8]

And the neighbors had more incriminating stories to tell. At all hours shouts and screams could be heard from the Mauroy home as the couple fought bitterly. But now Mauroy was not the only one to become violent during their disputes. According to these witnesses, Perrine "gave [her husband] many strokes . . . [and] having once received a box on the ear from him, she said, he should repent it, or he should die on it."[9] The neighbors testified that Perrine made good on her promises. They claimed that she "made a show of tasting [her broths] herself" to reassure her increasingly suspicious husband that his food was safe. Yet on more than one occasion, one witness had seen her cast "down on upon the ground, what she had in the spoon."[10] What was more, neighbors had seen "certain Powders" in Perrine's home, powders that—they were sure—were poisonous. Later in the trial Denis also testified that he had heard Mauroy scream in terror when he and Emmerez had arrived at his home to perform the third transfusion. The man howled with panic that his wife had been making plans to get rid of him.

On the face of things, the evidence against Perrine Mauroy was meager and based only on hearsay. One is also left to wonder why an otherwise shrewd enough woman would have been so public with her animosities, at least according to the neighbors. Moreover, if she had been part of the higher classes to whom she was now pleading her case, the evidence would have not been enough to implicate her directly in what was increasingly looking like murder by poison. Yet justice in seventeenth-century France was not blind, particularly not in the courts of the Châtelet, where judges had seen and heard every crime imaginable among the populace. Perrine was not only less articulate and had fewer resources than Denis, she had also been accused of a crime that the king's police chief, Nicolas de la Reynie—and by extension his judges—could not have loathed more: poison.

As the Duc de Saint-Simon, the celebrated memoirist of Louis XIV's reign, remarked: "It seems that there are, at certain moments, crimes which become the fashion, like clothes. Poisoning was *à la mode* at this time."[11] The most famous yet most elusive figure in the lethal world of poison was Catherine Deshayes. Operating under the name La Voisin, Deshayes lived in the village of Villeneuve, on the outskirts of Paris at the dead end of the abysmal rue Saint-Denis. For decades she had opened her home to a parade of women seeking assistance in seducing love interests, removing rivals from their paths, or eventually snuffing out a lover, a husband, or a wealthy relative. There was no adulterous condition that an "inheritance powder" or "soup from Saint Denis" could not cure.[12]

Despite Reynie's brutal attempts to crack down on criminal activities in Paris, the police chief had frustratingly little control over La Voisin's network of sorcery and poison. While Perrine

Mauroy was supposedly preparing deadly powders for her husband, the king's premier mistress herself was making her first visit to La Voisin. Between 1666 and La Voisin's arrest in 1679, the Marquise de Montespan regularly employed the sorceress's services to cast love spells and to create custom love potions so that she might retain the king's ever-fickle affections. After three years of secret court hearings, La Voisin was publicly beheaded in 1682—but not before more than four hundred people, many from the highest levels of Parisian society, had been accused of dabbling in poison. The Marquise de Montespan was among them. She was accused of sticking pins in a wax doll to punish Louis XIV for his affairs, and even of discussing with La Voisin the possibility of poisoning him. Louis was said to have been devastated by this news and other stories of evil in his court, but Montespan was never punished formally for her transgressions. She spent the last years of her life in self-exile at a convent in the provinces, far from the probing and accusatory eyes of the court.

The Affair of the Poisons—as this sordid case was eventually called—had more than convincingly shown how easily death could be orchestrated in late-seventeenth-century France. Rich and poor alike had ready access to local "sorceresses" of varying social standing who created secret and deadly herbal distillations containing mandrake, ergot, opium, juniper, and other substances. According to existing documents related to the Denis trial, the widow Mauroy had chosen a less exotic way to dispose of her husband. Perrine gave her husband broths loaded with arsenic, which had the advantage of killing its victims slowly and imperceptibly. Readily available in apothecary shops for use as rat poison, the symptoms of arsenic poisoning were indistinguishable from those of other maladies that regularly haunted early Europeans: cholera, dysentery, and plague.

Denis and Emmerez had made no note of diarrhea and vomiting during their last fateful visit to the Mauroy home. However, both said that Mauroy had looked more emaciated than usual and, of course, had exhibited intensely delusional behavior. Furthermore, police investigators claimed that a cat had died not long after having been administered "powders" from a vial they had found at the Mauroys' while collecting evidence against Perrine. However, the most damning evidence of all was Mauroy's mental state itself. Arsenic is harmful to the nervous system, and common symptoms of arsenic poisoning include delirium, tremors—and seizures.

Having heard all the testimony against Perrine, Judge Defita exonerated Denis of all accusations that his transfusions had killed Mauroy. The widow was formally charged with murder and promptly taken away. Beaten and abused as Perrine had been during most of her marriage, this fact was not taken into account in the judge's ruling. From this point forward she disappears from the historical record. This leaves us to assume that she spent her last days—however many there were—in one of the Grand Châtelet's horrific prison cells. Life had not been kind to Perrine, and early modern justice would do her no favors.

Although Judge Defita had declared Perrine fully responsible for her husband's death, he also confidently stated his belief that she likely had help. Perrine seemed an improbable candidate for such a well-thought-out plan. Moreover it was clear that she barely had the resources to take care of basic needs, much less purchase poison. As much evidence as had been presented against Perrine Mauroy, some key aspects of the case remained unclear. Defita asked three questions: Where did the widow get the powders that the neighbors had seen? Why did she give them to her husband? And most important of all, by whose suggestion?

Denis had stated in his initial complaint and in the trial at the

Châtelet that Perrine had tried to blackmail him. The widow admitted that some physicians had visited her in the days following her husband's death and had offered her a large amount of money in return for filing a murder case against Denis. Likewise, during the trial one of the neighbors testified that a Paris physician had also visited his home and offered him twelve pieces of gold "if he would depose that Mauroy died in the very act of the transfusion."[13] While all the witnesses made reference to these mysterious physicians at some point during the trial, the conspirators' identities were never formally revealed in the court proceedings.

Most people familiar with the case or the controversies surrounding transfusion in France in the 1660s would have had their own suspicions regarding the identities of the "Enemies of the Experiment," as Denis called them. And Denis himself may have whispered their names to Commissioner Le Cerf when he made his initial complaint. In the end Defita declared that the matter of Perrine's accomplices was "important enough to inquire into the bottom of it." For the unnamed physician and others who "solicited her with money to prosecute those that had made the operation and who had been seen with her," a day would soon be set to appear personally before the Criminal Lieutenant.[14]

Denis breathed a sigh of relief. The court had sided in his favor, and it seemed likely that Perrine's accomplices would be brought to justice. But he could not have been prepared for what followed next. Defita concluded his judgment by stating unequivocally that, from this point forward, "no transfusion should be made upon any human body but by the approbation of the physicians of the Parisian Faculty [of Medicine]."[15] The irony of the situation was clear to Denis and to all: The judge had just placed the future of transfusion in the hands of men who would never again allow it to be performed.

Denis refused to let the matter drop. He remained convinced

of the usefulness of transfusion—and especially of his right to perform experiments to perfect it. Soon Denis was making the rounds of various members of the medical school. He was not only pleading his case but collecting signatures. One month later he was proud to report that seven or eight physicians out of a body of over a hundred had signed.[16]

When it became clear that his petition would be insufficient to bypass the Châtelet ruling, Denis turned once again to his benefactor, Henri de Montmor. The nobleman's reputation in the scientific community and in social circles had been tarnished in the wake of the new Academy of Sciences, but his connections in the legal world of late-seventeenth-century France remained strong. As a lawyer who had held the title of Master of Requests in parliament, Montmor understood French judicial procedures and knew where strategic opportunities were to be found.[17] And, just as important, the nobleman had the means to foot Denis' ever-mounting legal bills.

With Montmor's help Denis appealed Châtelet's de facto ruling against transfusion to parliament. Sitting across the river from the Châtelet on the Île de la Cité, parliament was the country's highest and most prestigious legal body. In contrast to the Grand Châtelet, where citizens of any ilk could bring their complaints to commissioners and where the city's worst criminals were brought to die, the proceedings of parliament were largely *huis-clos* (closed-door) affairs. It served as a supreme court of sorts for high-profile criminal cases, which were tried in the turreted Chambre de la Tournelle.

While there was certainly a criminal component to his case, Denis and his lawyers instead brought their appeal to the Grand'Chambre, which took on all cases involving corporate bodies such as hospitals, guilds, and universities. In doing so the transfusionist had made clear his intention to take on the

Faculty of Medicine at the University of Paris. Parliament had shown itself willing to go against the medical school a year earlier when it declared—against the strong resistance of the faculty—that antimony and other chemical remedies were permissible in medical practice. Denis remained hopeful that the parliament would stand up once again, and rule that transfusion need not fall within the jurisdiction of the intensely traditionalist school.

Broadsides were plastered on the walls of buildings in the Latin Quarter and throughout the main thoroughfares of Paris. These large, cheaply printed posters had once shared the extraordinary news of the madman Mauroy's remarkable cure by transfusion.[18] Now they were covered with new ones announcing the upcoming hearing at parliament. Blood transfusion was on trial, they declared. And all of Paris waited eagerly to hear what the verdict would be.

The trial that took place on November 28, 1669, was closed to the public, but it was by no means a small affair. Well-dressed coachmen navigated elaborately decorated carriages through the large gates of the parliament complex as they delivered some of the highest-ranking men of French society to the illustrious Grand'Chambre. Crown princes and dukes joined the archbishop of Paris as ex officio members of the court. They all settled into the richly upholstered chairs that awaited them. And a "world of other great persons, men and women" murmured to each other eagerly as they waited for the show to begin.[19]

If the courtroom's gold-leaf mirrors, its embellished walls, and heavy drapery looked fit for a king, it was because they were. It was here in the Grand'Chambre that the French kings had held critical *lits de justice* (literally, beds of justice). The king traveled from his stately home at the Louvre to meet formally with parliament when he did not see eye to eye with its leaders there. Since the 1661 trial of Nicolas Fouquet and what amounted to the unceremonious dismissal of the legendary *parlementaire* Olivier Lefèvre d'Ormesson, the judges and lawyers in this highest French court had become decidedly more docile. Official meetings between the king and the courts became fewer and farther between as the young Louis XIV successfully consolidated his power as Sun King: No one dared go against him, and he would soon have no further need for the *lits de justice*.[20]

Denis should have known that transfusion stood no chance in the king's courts. But the line between an overabundance of hope and quixotic delusion is rarely clear. Throughout the course of his career, the transfusionist failed to recognize the extent to which the culture of medicine and science in the seventeenth century was organized according to a strict hierarchy built around money, power, and reputation. As an outsider by both birth and training, Denis also failed to appreciate how unyielding this closed world was. He delighted in breaking its rules and was convinced that, one day, he would be accepted in the inner sanctum on his own terms, in his own way, and entirely for who he was. He could not have been more wrong.

Transfusion was, of course, hardly the best choice of vehicle to reach such lofty goals. The very idea of transfusion flowed from theories and practices that, in Catholic France, were simply untenable. William Harvey's discovery of circulation in the 1620s had set off a critical rethinking of medical philosophies that had endured since antiquity—and in many French circles these traditional modes of understanding the body still dominated. Later infusion and transfusion experiments relied as well on Cartesian theory, which had also been received with no small amount of prickliness among Parisian scientific, political, and religious elites. Descartes' dualist philosophy had emphasized the division between body and soul. Notions that animal and human bodies were elaborate machines had been useful in scientific circles—not least in regard to providing new opportunities to perform research on live animals. Yet such arguments of a body-soul divide flew in the face of the most important traditional teaching of all: the Bible. To imagine transfusion meant to dismiss biblical dictates such as in Deuteronomy 12:23, "Eat not the blood, for the blood *is* the life." And for the French, the fact that the concepts of blood circulation and transfusion originated across the Channel in the camps of their heretical Protestant enemies the English only confirmed their resistance.

Denis' attempts to build a reputation on the new science of transfusion were clearly doomed from the beginning. The parliament's verdict on transfusion was similarly predetermined. Nonetheless Denis' appeal was given full consideration—at least in appearance. Arguing on behalf of Denis and transfusion was none other than Chrétien de Lamoignon, son of the highest-ranking member of parliament, the Honorable Guillaume de Lamoignon. Working from the investigation reports and testimony from the Châtelet hearing, the young Lamoignon made the case for Denis, and transfusion more generally, before the presiding judge, Matthieu Molé. Judge Molé was himself the son of the man who had preceded Lamoignon's father as first president of parliament. In other words, Denis' case as well as that of transfusion more generally was heard not only in the highest court in France but also by some of its most illustrious members. According to the short and only report we have of the hearing, Lamoignon's arguments to Molé were said to have been a "masterpiece."[21] Yet during the hearing surprisingly few questions were asked about the details of any possible plot against Denis, and there was no discussion at all of the potential identity of a man or men who helped Perrine hasten her husband's death. Clearly the fate of transfusion was the sole issue at stake in this trial.

The verdict came swiftly. From his high-backed chair on the court's central podium, the judge demanded silence. He stared dispassionately at the nobles, physicians, and lawyers who returned the gaze with respect and anticipation. We do not know what Denis' thoughts were as he waited for Molé to make his proclamation. But if he was not yet fully humbled by this unequivocable show of the court's authority—and, by extension, the king's power—he soon would be. Molé declared that he saw absolutely no reason to overturn the Châtelet decision. Blood transfusion would now, and always, be performed only with the express approval of the Paris Faculty of Medicine. And like everyone else in attendance, Denis knew the faculty had no intention of ever allowing that to occur. The courts had spoken not once but twice. Transfusion was officially dead.

Chapter 16

CHIMERAS

The Denis affair spelled the de facto end of transfusion not only in France but also in England. There is no evidence that the English officially banned the procedure.[1] Yet there is little doubt that the declaration at the French parliament chilled transfusion research across the Channel and throughout Europe. Members of the Royal Society turned their focus on blood's other mysteries, such as its chemical properties, clotting mechanisms, and the source of its florid color. Rethinking the long-held belief that the heart's heat gave blood its color, Richard Lower performed simple experiments to show that arterial blood still remained red even when cooled, and that venous blood changed color when exposed to air. To prove this he exposed the trachea of a dog and capped it off. Soon the blood flowing in the arteries was "completely venous and dark in color."[2] When the dog died, he pushed the venous blood through the dog's lungs, which had been perforated during the procedure. The blood turned bright red. Lower's questions on the importance of air and the lungs in blood set off a flurry of research over the years that followed. Depending on one's view, this new direction either eclipsed the English fascination with transfusion or served as a welcome alternative to a procedure that was now more cloaked in controversy than ever before. As for France, no transfusions were performed there until the nineteenth century, and blood science once again took a backseat to mathematics, physics, and astronomy.

While most historians have left the story there, an important mystery still remains. Who helped Perrine Mauroy poison her husband? And why?

The answer can be found in a single letter that has long sat unnoticed in the archives. The seven-page document is titled simply "Reflections by Louis de Basril, Lawyer in Parliament, on Disputes Concerning Transfusion." The document lacks a publisher's name, publication date, and documentation of royal permissions (*privilège*) normally required for publication. However, we do know that it was published before the trial at Parliament and possibly the one at Châtelet; a print excerpt of the letter was included in a 1668 collection, "Some New Observations on the Very Considerable Effects of Blood Transfusion." The print run for the full-length letter itself was very small; only four libraries in the world own copies. All of this suggests that Basril's letter circulated as a loose-leaf pamphlet, perhaps surreptitiously. And for good reason. As Basril makes clear in his letter, transfusion debates had taken a hostile and dangerous turn. Everyone was looking over their shoulders, it would appear. If they were not, they should have.

Still, the otherwise little-known Basril felt compelled to speak out. He explained that his reasons for publicly revealing the identities of those behind the plot against Denis were not motivated by a wish to argue either for or against the procedure. Instead he believed "with all of his heart in the truth"—and it was only through experimentation that such truths could be found. "In effect," Basril explained, "because transfusion is the subject of so many disputes and animosities, it seems to me that, to proceed sincerely, those who have declared themselves against it would do better to perform experiments and to examine it in good faith."

Details of Basril's life and status at parliament have been lost to history, but his words have not. In his revelatory letter Basril spoke calmly but firmly about the "indignation" that he felt toward those who "by ignorance or jealousy" worked to put an end to those experiments. Bristling over what he called the "cabal" against Denis, he named two men—Guillaume Lamy and Henri-Martin de la Martinière—for their involvement in the "secret intrigues" and "cowardly plots."

While Martinière and Lamy's paths had not likely crossed before Denis' experiments, the two men shared a firm bond from the moment the transfusionist began his animal-to-human experiments. A footman had presented Martinière with a letter by Lamy denouncing transfusion immediately following a vivid dream about a philosopher who was transformed into a cow. And from that moment, Martinière shared a "friendship [with Lamy] that I have imprinted in my soul."[3]

There is little doubt that Lamy had been working assiduously to turn the opinion of the Paris Faculty of Medicine against Denis and transfusion. Barely days after Mauroy's death, on February 16, 1668, Lamy wrote another letter to the influential physician René Moreau. His tone was reflective and measured, but Lamy could not resist gloating over Denis' now-uncertain future. "I would like to think," he crowed, "that . . . [Denis] saw in his imagination his reputation soaring and that all knowledgeable men would be praising his glory and miracles. But the human condition is subject to prompt change and marvelous vicissitudes. The miserable adventure of the madman's death will be enough to overturn all of his beautiful imaginations and to ruin entirely his high hopes."[4]

Lamy acknowledged in his letter that he and Denis had traded angry words in public. The University of Paris faculty member also acknowledged that there were lingering doubts among those in the medical world about whether he might have attempted to seek "vengeance against Monsieur Denis for having treated me so outrageously."[5] In the wake of such accusations Lamy resolved publicly to abstain from any public discussion about the matter from this point forward. "I do not wish to debate this issue further with him. I will not write of it again, not out of fear of accusations but for my good rest and also because I think that I have said enough."[6] Lamy kept his resolution until his death, it appears. And in the absence of other documents that would support Basril's accusations, it is difficult to say with confidence what Lamy's precise role may have been in assisting Perrine Mauroy in her dark deeds.

But there is no lack of historical evidence in the case of Martinière. While Lamy remained quiet, Martinière spilled copious amounts of ink proclaiming his innocence. But a paper trail followed him, and it implicated him directly in the death of Mauroy. As a devout Catholic, Martinière made no secret of his belief that transfusion corrupted both bodies and souls. A man with a decidedly colorful history, Martinière agreed wholeheartedly with the Paris Faculty of Medicine and the Academy of Sciences about the pernicious effects of transfusion. However, he was furious that Denis' research had been allowed to continue seemingly unchecked in the days and weeks preceding the Mauroy transfusions. Verbose and prone to passionate outbursts. Martinière fired off countless letters to any person he thought might listen.

For as powerful as Louis XIV had become, the pirate-turned-physician felt that Denis needed to answer to still another power—one even more commanding than the Sun King himself: God. Any notion of an immaterial soul, he argued venomously, was "ridiculous" and sacrilegious.[7] He shared this view with Lamy. Similarly rejecting Cartesian mind-body dualism, Lamy believed, like most of his colleagues at the university, that the human soul was corporeal.[8] Martinière went one step further, arguing that blood was the precious fluid that created a "harmonious link between the soul and the body."[9] For Martinière both transfusionists and Protestant alchemists were cut from the same cloth. Each tried to bring about unholy transformations. Alchemists transmuted metals. Transfusionists transmuted souls.[10]

Like the fears of hybrid monsters that Martinière nursed since his pirate days, so too did his scorn of alchemists come from personal experience. After his release from captivity on the corsair ships, he made his way through Portugal and Italy before ending his travels in France. Along the way he sojourned for two months in Milan, where he earned his keep by working for an alchemist. His job was suffocatingly hot and backbreaking. As a *souffleur* (puffer) he kept the fires burning as his master tried desperately to unlock chemical arcana. He watched his master's greedy quest for wealth, power, and eternal life—and then he snapped. "After having puffed three days and three nights with someone without accomplishing anything but wasting our time, our fuel and all of our lead, I took a bat," he explained, "and smashed all of the furnaces, kettles, alembics, and cauldrons. I swore that I would never again pursue the mad search for the so-called Philosopher's Stone."[11]

Martinière wrote a series of similar battle cries against transfusion between Mauroy's first two procedures in December 1667 and the fatal third one during the week of February 15, 1668. He described dreams in which he saw himself taking on with courageous pride and violence the monsters created by transfusionist

FIGURE 22: An alchemical laboratory. Young men were frequently employed as "puffers" (*souffleurs*), responsible for keeping fires burning underneath the alchemist's experiments. Phillip Galle (sixteenth century).

transformations. In one he claimed to have seen a Chimera, a monstrous beast with "the head of lion, the tail of a dragon, the stomach of a goat" along with other composite parts of beasts and humans. The monster "infected all parts of the earth where it roamed spewing pernicious venom." Transfusion, he explained, was Satan's work and solely responsible for resurrecting this and mythical monsters. "I believed," Martinière wrote, "that time had buried [them] . . . but Satan, enemy of the human race, on the pretext of charity, reignited them through vain hopes of [transfusion's] usefulness."[12]

In the dream Martinière readied his weapons to strike the beast. As he did, "a quantity of learned men" surrounded him.

They menaced him and threatened him until he ran away in surrender, fearful for his life. Moments later, the physician related, Athena herself—goddess of wisdom—arrived. Holding a large javelin in her hand, she impaled the monster and then beat it violently with a club until it died. Reflecting on his dream, Martinière vowed never to run away from his responsibilities again. He would stare the transfusionists straight in the eye. And then he would destroy them.

Over the nights that followed Martinière claimed to have been visited again by another mythic personage. This time he had a vision of Medea, the murderous transfusionist of antiquity. "I saw in the sky a woman in a chariot," Martinière claimed. "She looked at me with angry eyes and hissed, 'If my charms are not strong enough to convince you to give up your resolve to abolish transfusion, I will rip you apart [as I did] my brother and the children that I had with Jason.' " The physician turned indignantly to the sorceress. "It is you, execrable Medea, deadly witch!" he cried bitterly. "Despite your threats, I promise you that I will never give up on my dedication to the public good. I will do everything in my power to expose your dark plans to the world."[13]

Taking a page from Medea's own playbook, Martinière made good on his promise. And so convinced was he of the righteousness of his cause that he made little attempt to hide many clues that implicated him in Mauroy's murder. In a treatise he wrote on April 4, 1668, not long after the Châtelet hearing, Martinière confessed that he met at least once—when or where he does not say—with Perrine to discuss the transfusions that had been performed on her husband. He also confirmed that he encouraged Perrine to consider filing a formal complaint against Denis, although he wisely neglected to mention whether any money had changed hands in the process.[14] And tellingly, in the mass of documents related to Denis' work, it is Martinière alone who provided the name of the apothecary, a Monsieur Claquenelle, from whom Perrine reportedly bought the ingredients for the powders she administered to her husband.

Thus it seems highly likely that Martinière counseled Perrine on what poison to use, how to acquire it, and how it should be administered. To be sure, Martinière was no stranger to the effects of various herbs and their poisonous potential. His first task in every port of call, as the sole doctor on a pirate ship, had been to seek out apothecaries. He had also published a lengthy *Treatise on Antidotes* just a few years earlier—and to know antidotes one needed to know poisons. The bulk of Martinière's treatise focused on the complex preparation of "mithridate," which many herbalists and apothecaries claimed to be something of a magical cure-all. Martinière's own recipe for mithridate contained more than forty different substances—including rose leaves, myrrh, and powdered extract of beaver tail glands.

The mixture was named after the second-century BC Greek king Mithradates VI, who was rumored to have hardened himself against poison by means of a mysterious and virtuous potion that the king researched throughout his reign. This antidote—and his obsession—earned him the title "poison king." He was infamous for testing his poisons, as well as his antidotes, on prisoners. At elaborate banquets prisoners condemned to death would be publicly fed poison-laced food or shot with poison-dipped arrows while the king narrated their symptoms to the crowd. When death was near, the prisoners were dragged away and used as guinea pigs to test the king's antidotes.[15] Martinière likewise observed without hesitation that poison, death, and doctoring went hand in hand. "I know," he argued, "that homicide by doctors is allowed."[16]

Martinière's writings continued to take on a dark and menacing tone. Martinière depicted himself as a noble warrior and left no doubt that his next target would be Denis himself. "Allow me to tell you, Sir," Martinière wrote directly to Denis, "that Satan reveals himself through your work."[17] He made it clear that he viewed himself as "a spirit that will not fail to arrive at our goals, by giving [the transfusionist] a fatal strike."[18] Mauroy's death, he explained, would be only a prelude to Denis' own: "We read in the book of Judges that the King of the Canaanites, who killed several kings by cutting their hands and feet off, was sentenced to die of their same death. And it is written in the New Testament that whoever kills by the sword, dies by the sword. Be careful," Martinière warned Denis, "that . . . you are not yourself visited by the Furies, who as principal guardians of the law, will perform endless transfusions on you with the help of their little minions from the underworld. Or [perhaps you should be] careful that you're not transmutated into a calf [as] Lucian was in the Golden Ass, sentenced to hard labor and beaten with sticks."[19] Martinière's ravings were not, it would appear, just the innocuous outbursts of a man who believed deeply in the righteousness of his claims. As the lawyer Basril noted, "Monsieur Denis is very prudent to keep silent."

Martinière was outraged when he became aware of Basril's letter naming him in the plot. "This," he wrote indignantly, "is my reward for having worked to suffocate the transfusionist monster in his cradle"—a phrase that Lamy, coincidentally, used as well in one of his own letters.[20] The angry doctor quickly went on the defensive and, in doing so, spilled yet more ink as he sent handfuls of letters far and wide in order to proclaim his innocence. Among these letters was one addressed directly to Mathieu Molé, the man assigned to judge Denis' appeal in parliament. Martinière urged the judge to take a stance against transfusion's unholy corruption of the human race. "Knowing that you are the Judge of the Transfusors," wrote Martinière, "I have taken the liberty to put my hand to plume in order to show you the horror of the operation, which is directly to the contrary and opposite of God's wishes, because it destroys His living images."[21]

Martinière continued his writing campaign at the highest of political echelons—going so far as to address Prime Minister Jean-Baptiste Colbert himself. "Knowing that you do not approve of this so-called operation of blood transfusion," Martinière declared, "I know that you will not be surprised to hear that there are some men who are so feeble minded to believe that it is an effective remedy against a variety of illnesses."[22] Martinière warned Colbert of the bloodbaths that were sure to happen should the transfusionist be allowed to continue his work. "Denis," Martinière explained, "is hoping that the doctors of the Faculty of Medicine will allow the deaths of seven to eight million men before they bring themselves to condemn the procedure."[23] Martinière begged Colbert to step in and offered a suggestion for a suitable punishment for Denis: "The innovators whose inclination is to pull and push blood . . . should be sent to the Caribbean and sacrificed to the idols."[24]

In his letter to the prime minister, Martinière reiterated his belief that his actions had been fully warranted and asked for Colbert's help against those who would seek to punish him for his action. Placing his fate in the prime minister's hands, he pleaded: "Your Goodness, I finish here by asking for the honor of your protection and by begging you to stop the case [against me], because I never intended to hurt anyone, I only wanted to ensure that this cruel and disgusting science would not endure."[25]

There is no formal record of Colbert stepping in, yet one thing is certain: Martinière was never put on trial. And without another word Martinière disappeared. In the years following his involvement in the Denis affair, he quietly published two lengthy memoirs of his travels as a young man on the high seas. Yet not a single one of his works addressed medical issues, nor did he ever discuss transfusion again. Between the parliament ruling and his death in 1676, biographers have found only fleeting traces of Martinière in Amsterdam and Dublin.[26] This leaves us to question whether there could have been a gentleman's agreement of some kind that recognized his usefulness in putting transfusion to rest, while at the same time removing the uncontrollable and explosive Martinière from the Paris medical community.

Denis' supporter Henri de Montmor also slipped from sight following the trial. According to Jean Chapelain, once a regular attendee of the Montmor Academy, the nobleman fell into a deep depression after 1669—the year of the parliament trial. "He was," wrote Chapelain, "obliged to sell his title as Master of Requests [at parliament] and there was some talk that he lost his mind a bit, or even fell into suicidal despair. For eight days, they had to force him to eat to stay alive."[27] It took a visit from the archbishop of Paris to persuade Montmor to let his family and his doctors care for him. And for at least a year after the trial, "he lived only on milk and did not involve himself in any domestic matters, nor did he speak to anyone or accept visitors." Giving up his hopes of a private academy for good, Montmor spent the next six years removed from the world in a state of disillusioned stupor.[28] Meanwhile his eldest son handled all family affairs, and soon the once-legendary family was bankrupt. Montmor died in 1679.[29]

As for Denis, he returned to his home on the Left Bank, where he gave paid lectures to students as he had done before transfusion catapulted him into the public eye. The experience had changed him, it seemed—or at the very least had taught him to temper his ambitions. One of his first conferences after the parliament hearing focused on "judicial astrology," or the ability to tell the future from the stars. "There is nothing so common as to see people who are infatuated with the folly of astrology, who brag about being able to predict various events that will happen in their lives," Denis explained, with some disappointment and perhaps newfound wisdom in the wake of his losses. "Anyway, if a prediction that good things will happen does actually come true, it would be a disservice. Predictions will always keep you in a state of suspense, in a state of impatient hope, and this hope will deprive you of everything that is good and agreeable in life."[30]

Yet Denis' endeavors later in life may just prove to be one of the greatest of history's frequent ironies. Four years after the final trial at parliament, the former transfusionist set himself on the most unlikely of research paths. Denis—the man who boldly championed transfusion against all odds—invented styptic, which is now found in medical cabinets around the world and used to stop mild bleeding.[31] If he was not able to ensure his legacy by making blood flow, he would do it by making blood stop entirely. Denis died in 1704 at the age of sixty-nine.

Another one hundred and fifty years would pass before blood transfusion returned to the early-medical landscape. In late 1817, James Blundell walked briskly down the halls of the maternity ward at Guy's Hospital in London. The twenty-six-year-old doctor had been called to the bedside of a new mother who was hemorrhaging. By the time Blundell arrived, the bleeding had stopped but the woman was pale and gravely weak from blood loss. With the angst of a newly minted physician who had not yet seen his fill of death, Blundell lamented that "her fate was decided." He was right; the woman died two hours later.

Years later, Blundell described his thoughts as he witnessed this "melancholy scene." He was haunted by a single question: Could he have done something to save the new mother? Blundell reached into the past and decided it was time to reconsider transfusion and to give this "neglected operation" the "experimental investigation which it seems to deserve."[32] "After floodings [hemorrhaging]," he wrote explaining his decision, "women sometimes die in a moment, but more frequently in a gradual moment; and over the victim, death shakes his dart, and to you she stretches out her helpful hands for the assistance which you cannot give. . . . I have seen a woman dying for two or three hours together, convinced in my own mind that no known remedy could save her: the sight of these moving cases led me to transfusion."[33]

Like Denis a century earlier, Blundell performed interspecies transfusions to test his theory. Using a syringe, he injected three dogs with human blood. All three died.[34] Several experiments later Blundell surmised that the "blood of one class of animals cannot be substituted . . . for that of another with impunity."[35] Turning his efforts to human-to-human transfusions, he soon recruited husbands as well as male hospital workers to serve as ready blood donors for mothers in need. The results were mixed. Of the ten patients he transfused over eleven years, only five survived.[36]

Blundell's work set off a rush of blood experimentation in the mid-1800s that culminated in Karl Landsteiner's landmark discovery of the ABO groups in 1901. In the wake of blood typing, blood transfusion quickly entered clinical practice. At New York City's Mount Sinai Hospital, for example, nearly fifty transfusions a year were performed between 1907 and 1914.[37] Still, transfusion remained a labor-intensive and dangerous procedure. Patients were required to be in the same room together, and a large staff was needed at the bedside: a surgeon, a surgical nurse, and nurses for each of the patients. Further, the donor risked long-term damage or even amputation of the limb if blood flow to the transfusion site could not be reestablished.

Fresh blood will begin to clot in just five minutes outside the human body. So any possibility of transfusion without the need for a donor at the bedside depended on finding an efficient and safe way to store and transport blood. Over the course of the nineteenth century, researchers experimented with a wide range of substances in the blood to combat clotting. At the turn of the twentieth century, just four years after Landsteiner's landmark discovery, researchers in three different countries, working independently of one another, recognized that the addition of sodium citrate to donor blood kept it from clotting. And unlike other substances tried in the past, it did not appear to have adverse effects for the recipient.[38]

It was on Europe's battlefields that blood transfusion effectively pushed its way to the forefront of medical practice. When the Spanish civil war broke out in the summer of 1936, bombings of major urban areas left thousands of civilians dead or gravely injured. Led by the physician Frederic Duran-Jordá, the Republican army health service set up a blood distribution network based on voluntary donors, who were called in by rotation at one-month intervals. Over the course of thirty months the Barcelona blood transfusion service recruited thirty thousand donors who provided more than nine thousand liters of blood used in twenty-seven thousand transfusions.[39] The blood was stored in five-hundred-milliliter reusable Erlenmeyer flasks that were sterilized by steam. Once the blood was warmed on-site, transfusions could be carried out on the field without further delay.

At about the same time, across the Atlantic, the first blood banks were being created. In 1937 Dr. Bernard Fantus established a central depot at Cook County Hospital in Chicago, which he later called a "blood bank," as a place where donors could have blood drawn and stored for future use, either by themselves or family members with the same blood type. When World War II broke out just a few years later, the American Red Cross helped organize a civilian blood service to support war efforts. The Red Cross opened its first center in New York in January 1941, and by the time the war ended in 1945, it had collected well over 13 million units (a unit is just under a pint) of blood.[40] The American Red Cross formally established a civilian blood service in the years following the war to help meet what had now become a soaring demand for blood in the States. And in November 1947, directors of more than fifty independent blood banks met in Dallas to draft the charter of the American Association of Blood Banks (AABB), which was established to ensure consistent, research-based standards for transfusion and whose work continues today with more than two thousand member institutions.

Today blood transfusion is one of the most commonly performed medical procedures in the world. In the United States alone, about 15 million pints of whole blood are donated annually by more than 10 million people.[41] Transfusion has become a gold standard in treating a broad range of illnesses and injuries, from chronic anemia to blood loss from trauma and surgery—so much so that it is impossible to tally the number of lives that have been saved or improved by the procedure.

Epilogue

I first stumbled on the Denis case nearly a decade ago while preparing course notes for an undergraduate lecture on Harvey's discovery of blood circulation. It was odd and fascinating, but other research projects beckoned. Still, over the years that followed, I could never seem to get Denis and the fate of early transfusion out of my mind. On research trips to France and England for other topics, I found myself stealing a peek at anything that had to do with early blood science. On one corner of my desk I kept a growing stack of research documents—articles, reading notes, illustrations, copies of manuscripts, and scientific correspondence—related to every aspect of early blood work. And nearby on the floor sat an equally overflowing pile of papers about animals, monsters, and interspecies hybrids in early Europe. I would not learn about Lamy, Martinière, and the Basril letter until much later and understand finally how the two were linked, but I became increasingly certain that early animal-to-human transfusions were a case study for larger political struggles, religious controversies, and cutthroat ambitions during the late seventeenth century.

Still, it was not until January 31, 2006—after several years of research and while listening to George W. Bush's State of the Union address—that I knew the story of early blood transfusion not only should be told, it *had* to be told. In his address Bush called for "legislation to prohibit the most egregious abuses of medical research [including] creating human-animal hybrids." Bush's speech echoed a report by the President's Council on Bioethics two years earlier, in 2004, which argued for a congressional ban on animal-human embryonic stem cell research as a way to prevent "some adventurous or renegade researchers" from doing untold damage to the human species.[1]

History was repeating itself. Jean-Baptiste Denis was also seen by most as a dangerous renegade. Yet he was championing a medical procedure that we know today to be invaluable. Of course there is more than ample evidence to suggest that personal glory and fame were Denis' prime motivators. In fact he actually did precious little original science himself. For men like Boyle, however, scientific research—and particularly transfusion research—meant pursuing questions about the natural and human worlds that were necessarily difficult and unavoidably uncomfortable. In the seventeenth century blood transfusion hit at the heart of what it meant to be human—and what it meant not to be. To imagine early blood transfusion was to imagine a world where hybrid species not only existed but could even be created by science.

In 1666 Boyle asked "whether by frequently transfusing . . . the blood of some Animal into one of another Species, something further and more tending to some degrees of a change of Species, may be affected." He speculated that transfusion would provoke no change in the recipient. Still, science had a responsibility to do whatever was necessary to ensure a definitive answer to the question. As

Boyle explained, it was "worthwhile for satisfaction and curiosity to determine that point by Experiments." Yet as the resounding and nearly immediate bans following the Denis case suggest, there were clear limits to how far the already fluid borders of the animal and the human could be pushed in the late seventeenth century.

In the days and weeks that followed the State of the Union speech, I kept watch on how news outlets and the general public responded to discussions of animal-human chimeras created through stem cell research. It was clear that the president was tapping into deep societal fears surrounding genetic manipulation and, more specifically, scientific research that combined human and nonhuman genes, embryos, and embryonic stem cells. Story upon story appeared on television, on the Internet, and in newspapers about interspecies experiments that scientists had either already performed or were on the cusp of performing. News reports described hybrid creatures born of science, each more stunning than the next: sheep that received human blood-forming stem cells and now sported a liver with more than 40 percent of their cells derived from human cells,[2] "humsters" created during human sperm-viability tests using hamster eggs, "geeps" made of fused goat and sheep embryos.[3] It was supposedly only a matter of time until the first "humanzee" would make its first laboratory appearance.

Such extraordinary examples aside, interspecies research actually happens daily in scientific laboratories across the world—and in much less sensational ways. Researchers have long injected human cells into mice in order to gauge the effectiveness of a vaccine, and there is a multi-billion-dollar industry centered on creating designer mice ("knockout mice") that are prone to specific human illnesses like cystic fibrosis or chronic conditions like obesity. Such interspecies research has proven invaluable to developing new drugs and procedures that save human lives or improve quality of life. Case in point: In 2009, a few months after her son left the White House, Barbara Bush underwent successful open-heart surgery to replace her aortic valve; the replacement valve was from a pig.

Transgenic intersections such as these have given rise to little public outcry. However, cross-species experiments that transplant human neural stem cells into animal embryos or brain tissue are surrounded in controversy. At what point does a mouse brain cease to be mouselike? And at what point would such a chimeric creature take on the moral status, rights, and responsibilities conveyed to humans? The central challenge here is not one of science. In fact science still has very far to go before it could ever catch up with the fictions of Dr. Frankenstein's monster and Dr. Moreau's island.[4] Instead the possibility of scientifically created animal-human chimeras, in our own era and in those that have preceded ours, necessarily force society to address issues of species integrity, moral taboo, human and animal dignity, and what is "natural."[5] But most of all we are asked to come up with an answer for the thorniest question of all: What does it mean to be "human"?

The summer following his 2006 State of the Union address, Bush issued his first veto in the five years that he had been in office. The veto stopped Congress's efforts to lift funding restrictions on human embryonic stem cell (hESC) research. In a dramatic and, for many scientists, a welcome turn of events, President Barack Obama signed an executive order in March 2009 that removed preexisting presidential actions on hESC cell research, thereby allowing the National Institutes of Health (NIH) and other agencies to substantially increase funding streams to researchers. This was short-lasting. In August 2010, a federal court issued an injunction calling for an immediate stop to any hESC research activities. As I write, human embryonic stem cell research is now at a critical crossroads, as policymakers and the public weigh its fate.

I am left to wonder if the seventeenth century would still have halted transfusion experimentation had it known the degree to which future generations would come to depend on it. How many lives would have been saved—or lost—if blood research had been allowed to move forward, instead of being relegated to the footnotes of history for centuries? Every era, particularly one as deep in "Scientific Revolution" as our own, must necessarily confront some of the same time-worn debates about whether the contours of the human body, mind, and soul are as stable as we might like them to be. My greatest hope is that when historians tell our own story decades and centuries from now, they will be able to say that we thought these issues through well and addressed them with fearless curiosity.

BLOOD TRANSFUSION

A Chronology

February 1665 (England): Richard Lower conducts the first dog-to-dog transfusion experiments.

April 1665–February 1666: Great Plague of London.

September 2–5, 1666: Great Fire of London.

November 14, 1666: The Royal Society returns to animal transfusion experiments.

January 22–March 21, 1667 (France): Claude Perrault, Adrien Auzout, and Louis Gayant begin animal transfusion experiments on behalf of the French Academy of Sciences.

March 3, 1667: Jean-Baptiste Denis begins independent transfusion trials and conducts more than twenty canine experiments as well as mixtures across different species (cows-dogs; horses-goats).

June 15, 1667: First animal-to-human blood transfusion performed by Denis, who transfuses a fifteen-year-old boy with lamb's blood.

November 23, 1667 (England): Royal Society members Richard Lower and Edmund King, transfuse Arthur Coga with lamb's blood. Coga is transfused a second time on December 14, 1667.

December 19, 1667 (France): Denis transfuses the madman Mauroy with calf's blood. Two transfusions follow over the course of two weeks. Mauroy later dies.

April 16, 1668: Trial at the Châtelet for the death of Antoine Mauroy.

December 1669: French parliament officially bans transfusion.

1818 (England): James Blundell performs the first successful human-to-human blood transfusion.

1867 (England): Joseph Lister uses antiseptics to prevent infection in blood transfusions.

1901 (Austria): Karl Landsteiner discovers first three human blood groups (A, B, O). Blood type AB is discovered the following year.

1908 (France): Alexis Carrel develops a method to prevent clotting by stitching recipient and donor vessels together; he received the 1912 Nobel Prize in Physiology or Medicine for his work.

1914: Anticoagulant sodium citrate is developed, which allows for blood storage and facilitates transfusion.

1932 (Russia): Lenigrad hospital establishes first blood bank.

1937 (United States): First American blood bank is established at Cook County Hospital (Chicago). Over next two years blood banks are established in Cincinnati, Miami, New York, and San Francisco.

1947 (United States): American Association of Blood Banks is created.

Acknowledgments

I had the very good fortune of being surrounded by many supportive and enthusiastic friends and colleagues while researching and writing this book. First and foremost, my thanks go out to Christine Jones and Miranda Nesler, who read every word of the manuscript several times over (actually many more times than that) and were always as honest and blunt in their criticism as they were their encouragement. Miranda, an equestrian as well as an academic, also helped make sure that every detail rang true—even if it meant containing her frustration with me for clearly having no clue about the texture of horses' noses and tongues.

Faith Hamlin also believed in this book from the start. *Blood Work* could not have found a better home, and I owe it to Faith for helping to make it happen and for being a trusted source of advice about the intricacies of the publishing world. When I call her my book guru, I mean it. I am also grateful to Courtney Miller-Callihan for all of her help and encouragement.

At Norton, I won the lottery having Angela von der Lippe as

my editor. Angela has worked with many of the writers in the history of science and medicine that I admire most, and it showed in her smart and probing questions that always pushed me to revise deeply—and for the better. Laura Romain also understood where I was heading with the book even before it felt like I did and shepherded the manuscript deftly through the publication process. Thanks as well to everyone at Norton who helped dot the i's, get the words printed on the page, and help put the book into readers' hands: Don Rifkin, Sue Llewellyn, Erica Stern, Rebecca Carlisle, Jess Purcell, and Melissa Whitley.

The seeds of this book were sown while I was a fellow and codirector of the Robert Penn Warren Center for the Humanities interdisciplinary seminar on race and sexuality before 1700. Leah Marcus, David Wasserstein, Dyan Elliott, Lynn Ramey, Katie Crawford, Lynn Enterline, Carlos Jáuregui, and Jean Feerick could be counted on for their expertise and laughter around the seminar table; and Mona Frederick, Lacey Galbraith, and Galyn Martin provided a welcome space, administrative support, and great conversation during my year in residence at the Warren Center. I also am grateful to colleagues at Vanderbilt and well beyond for their advice and suggestions: Jay Clayton, Ellen Wright Clayton, Larry Churchill, Arleen Tuchman, Ed Friedman, Jérôme Brillaud, Matt Ramsey, Carl Johnson, Kendal Broadie, Mark Woefle, David Boyd, Michael Bess, Jeffrey Tlumak, Carlina de la Cova, Marri Knadle, Craig Koslofsky, Anthony Turner, Peter Mancall, David Kertzer, Jonathan Sawday, and Matthew Cobb.

Vanderbilt provost Richard McCarty and my department chair, Lynn Ramey, helped me find both the time and resources that were critical to my work on the book. My research during the early stages of the book benefited as well from the generous support of the Wellcome Library for the History of Medicine, the Newberry Library, and the Francis C. Wood Institute for the History of Medicine.

Behind every good research-based book is an even better librarian. And in this case, I have a large handful of librarians and library staff members to thank for their herculean efforts in helping me gain access to the texts and images I needed for this project. Mary Teloh, Jim Thweatt, Yvonne Boyer, and Jim Toplon at Vanderbilt University; Crystal Smith at the National Library of Medicine; Rachael Johnson at the Wellcome Library for the History of Medicine (London); Estelle Lambert and Bernadette Molitor at the Bibliothèque Inter-Universitaire de Médicine (Paris); Christiane Pavel at the Archives de l'Académie des Sciences; and Maria Conforti at the Università de Sapienzia (Rome). This book similarly could not have been written without the superb and important work of scholars in the field. These sources are listed in the notes and the bibliography. However, I would like to call out a few scholars who, *à leur insu*, were my surefooted companions on the journey: Harcourt Brown, Marie Boas Hall, Lisa Jardine, Robert G. Frank, Joan DeJean, Roy Porter, and Orest Ranum.

My research assistants, Jared Katz, Rob Watson, Megan Russell, Kiana Jansen, and Megan Moran, went above and beyond the call of duty in their work on the project. Louis Betty, Essie Assibu, Cate Stewart, and Pierre Hegay stepped in to help on the technical side of manuscript preparation when deadlines loomed; and Todd Dodson and Chris Noel took care of the inevitable computer breakdowns along the way.

Dr. John B. Breinig read the full manuscript with a physician's eyes; I see John, a talented photographer as well, when I catch a glimpse of the author photo on the book jacket. We weren't sure how long we had left with him at the time, which made the afternoon he and I spent taking the pictures all the more precious. Thanks also go out to Lauren Schmitzer, Mary Lawrence Breinig, and Janice Haithcoat, who offered feedback on various parts of the book. Liz Shadbolt and my mother, Carolyn Tucker, also read an early version of the book and ended up saying just the right things at just the right time. Their comments meant reworking what felt like the entire manuscript, but there is little doubt in my mind that the book is much, much better for their input. Al Hamscher also talked me down from the ledge after months of hitting my head against the intricacies of French parliamentary procedures and documents. I look forward to more martini-fueled conversations about our shared obsessions for seventeenth-century life.

Writing can be a solitary endeavor, but I've lucked out to be surrounded by good friends who help make it more bearable. Many thanks to Todd Peterson, Roberta Bell, Brooke Ackerly, Lauren Schmitzer, Anita Mahadevan-Jansen, Duco Jansen, Jeff and Rachel Haithcoat, Trish and Chris Juoza-Clark, and Susie and Marvin Quertermous, and the late Elinor Lykins. I'm grateful as well to a number of others who have helped and inspired me along the way: Tracy Barrett, Louhon Tucker, Delia Cabe, Maryn McKenna, Margaret Littman, Dan Ferber, Sandra Gulland, Michelle Moran, all of the FLX crew, Mark Evitts, Sandy Beckwith, Diane Saarinen, Lisa Morosky, Joanne Manaster, Katie Davis, and fellow volunteers at the Nashville Red Cross.

While researching this book, I traveled often to work in specialized library collections abroad and to retrace the steps of my transfusionists in both Paris and London. In many cases, the physical spaces are either no longer there or have been reconfigured in a way that makes their historical significance hard to appreciate fully. This was not at all the case at the Hôtel Montmor. I remember waiting for what seemed like hours one beautiful May day for someone to exit the soaring wood doors of the still private estate so I could steal a peek of what lay behind. Once I finally gained access, I stood in the *cour d'honneur*, mouth agape and eyes watering. It was precisely as I had imagined it after having worked so long with period documents, maps, and architectural plans. The concierge of the building, Jean-Marie Carpentier, approached me cautiously, likely wondering if I were not a little bit of a nut case. As I explained, the words came tumbling out in near gasps. Up *there*, that's where the Academy met. *This* is the staircase Denis walked up the night of his history-making experiment. Under these dormers *here* is where Mauroy stayed after the transfusion. After Monsieur Carpentier confirmed to my delight that the staircase, the balustrade, the tiles, all of it was in its original state, we spent the rest of the afternoon together exploring the building and what is left of the gardens—teaching each other about the rich history of the building. As we strolled among the ghosts, I was reminded once again of how insanely grateful I am to do the type of work I do as a researcher and a teacher.

And finally, two people in my life probably qualified for sainthood for their stoic patience, kindness, and love while I was in the depths of the book. My husband, Jon Hamilton, held down the fort during my many research trips and took on the one thousand tasks I overlooked even when I was home. Still, he found the time to make a model of Blundell's "Gravitator" blood transfusion device for me—out of Legos. Audrey, my grade-school daughter, kept me on task by having me sign weekly progress contracts. If there is one person I hope to never disappoint by breaking a promise, it is my daughter and especially one who proudly declares to anyone who will listen that her mother is a writer. This book is dedicated to you, Audrey. Heart and soul, breakfast, lunch, and dinner.

Notes

INTRODUCTION

1 Schmidt, "Transfuse George Washington!" 275–277. See also Morens, "Death of a President."

2 Thornton, *Papers*, vol. I, 425, 528–529. My emphasis.

3 Ibid.

4 Shapin, *The Scientific Revolution*, 3. Shapin begins his book with what may be my favorite quote ever on the topic: "There was no such thing as the Scientific Revolution, and this is a book about it" (1). If I have retained in some circumstances the upper case *S* and *R* in my references to the "scientific revolution" in the book and in the subtitle, it should be understood only as a way to reference a specific approach in these historiographical debates, rather than an affirmation of a singular worldview or uniformity in cultural practices.

5 Maluf, "History of Blood Transfusion," 67.

6 Brown, "Jean Denis and the Transfusion of Blood," 15; Farr, "The First Human Blood Transfusion," 151; Maluf, "History of Blood Transfusion," 66–67.

7 Pepys claimed that his stone was a big as a tennis ball. Writing on March 26, 1659, Pepys explained: "This day, it is two years since it pleased God

that I was cut for the stone at Mrs. Turner's in Salisbury Court. And did resolve while I live to keep it a festival, as I did the last year at my house."

8 Mauriceau, *Maladies*, 357.

9 For an outstanding analysis of blood and racial politics in the twentieth century, see Wailoo, *Drawing Blood*, 134–161.

10 Pete Jarman, 10 March 1945. Cited in Love, *One Blood*, 194. The writer of this letter neglects the fact that black soldiers also fought in World War II.

11 By the early 1950s serological research had successfully determined that Rh-factor sensitization as well as a variety of antibody phenotype groups were an important complicating factor when selecting appropriate donors for blood transfusions. This research also allowed for the possibility of using blood groups as genetic markers for specific diseases and for race-based "traits." For a detailed analysis of the controversies surrounding race in blood from the 1950s to the mid-1960s, see Kenny, "A Question of Blood, Race, and Politics," 469. Much of my discussion here is drawn from his work.

12 Scudder, "Sensitising Antigens as Factors in Blood Transfusions," 99.

13 Ibid., 99–100; Scudder and Wigle, "Safer Transfusions," 78–79.

14 "Seven at Columbia," *New York Times*, November 15, 1959, A50; W. L. Thurston, "New Procedure Advocated," ibid., E11. Cited in Kenny, "A Question of Blood, Race, and Politics," 462–463.

15 Giblett, "A Critique of the Theoretical Hazard of Inter. vs. Intra-Racial Transfusions," 233–277. Cited in Kenny, "A Question of Blood, Race, and Politics," 465.

16 For more on race and transfusion, see Kenny, "A Question of Blood, Race, and Politics"; Lederer, *Flesh and Blood*; and Wailoo, *Drawing Blood*.

17 Leshner, "Where Science Meets Society," 815.

Chapter 1: THE DOCTOR AND THE MADMAN

1 The 1660s were marked by what is now called the "Little Ice Age," which brought winter temperatures several degrees lower than in the present day. See Fagan, *The Little Ice Age*, and Brown, *Scientific Organiza-*

tions, 78. Writers such as Ismael Boulliau, a contemporary of Denis and Montmor, complained of a cold so intense that ink froze in inkpots.

2 Hussey, *Paris: The Secret History*, 140.

3 The Pont-au-Change as it stands in Paris today replaced its seventeenth-century predecessor in the late 1850s.

4 Hillairet, *Dictionnaire historique*, vol. 2, 303–304. For details on Parisian shops, see Blegny, *Livre Commode*, 238, 261, 192. Dueling was outlawed in 1602.

5 Peuméry, *Jean-Baptiste Denis*, 8–9.

6 Orest Ranum's *Paris in the Age of Absolutism* was an invaluable example and resource as I shaped my own descriptions of the sights, smells, and sounds of seventeenth-century Paris. See especially chaps. 1 and 6.

7 The street has since been renamed the rue du Temple; the estate remains, however, and has recently been restored to its original grandeur.

8 Collins, *The State in Early-Modern France*, xxvii.

9 Tallement des Réaux, *Historiettes*, 294–295.

10 Kerviler, "Henri-Louis Habert de Montmor," 199.

11 Ibid., 202.

12 For a description of Montmor's typically sanguine demeanor and generous hospitality, see Sorbière to Hobbes, early 1663 in Hobbes, *Correspondence*, vol. 2, 547. Cited in Sarasohn, "Who Was Then That Gentleman?" 219.

13 Emmerez (also spelled "Emerez") was praised as one of the best surgeons in France and was known for his careful and skillful work. He died on September 7, 1690. See Brown, "Jean Denis and Transfusion of Blood," 15, and Eloy, *Dictionnaire historique de la médecine*, vol. 2, 138.

14 All details of the transfusion, its circumstances, and its outcome are drawn from: Denis, "Cure of an Inveterate Phrensy" and the many other accounts published in the *Journal des sçavans* and the *Philosophical Transactions* as well as Poterie, Lamy, Martinière, and Oldenburg.

15 Poterie, Letter on transfusion, 28 December 1667, n.p.

16 Watkins, "ABO Blood Group System," 243.

17 Denis, "Cure of an Inveterate Phrensy," 622.

18 The irony of the priest's name—which means "young cow" or "veal" in French—went unnoticed by the self-absorbed Denis and other eyewitnesses to the experiments.

19 Denis, "Cure of an Inveterate Phrensy," 622.

Chapter 2: CIRCULATION

1 For a discussion of the early church attitudes toward dissection, see Park, "Myth 5: That the Medieval Church Prohibited Human Dissection," 43–49.

2 See Babington, "Newgate in the Eighteenth Century," 650–657.

3 See Linebaugh, "The Tyburn Riot Against the Surgeons," 65–118.

4 Harvey also dissected his wife's beloved parrot after its death. See Keele, *William Harvey*, 29, 51.

5 Claudii Galeni, *Opera omnia*, vol. 11, 149, 281.

6 Cited in Elmer, *Health, Disease and Society in Europe, 1500–1800*, 63.

7 Paré, *Workes*, 692.

8 Ibid.

9 Morabia, "P. C. A. Louis and the Birth of Clinical Epidemiology," 1330.

10 McMullen, "Anatomy of a Physiological Discovery," 492.

11 Boyle, *A Disquisition about the final causes of natural things*. See McMullen, "Anatomy of a Physiological Discovery," 493–494.

12 Harvey, *De motu cordis*, chap. 9.

13 Cited in Frank, *Harvey and the Oxford Physiologists*, 174.

14 Harvey, *De motu cordis*, 96.

15 British Library Add. ms 25071, folios 92–93. See Jardine, *On a Grander Scale*, 122–123, 511n51.

16 Cited in Wren, *Parentalia*, 62–63.

17 Ibid., 123. See also Frank, *Harvey and the Oxford Physiologists*, 171, and Gibson, "Bio-Medical Pursuits," 334.

18 Frank, *Harvey and the Oxford Physiologists*, 172.

19 Boyle, *Correspondence*, vol. 4, 357ff, and Frank, *Harvey and the Oxford Physiologists*, 172.

20 Oldenburg, *Correspondence*, 356, 366; Boyle, *Usefulnesse*, vol. 2, 64.

Chapter 3: THE AGE OF VIVISECTION

1 Descartes, *Discourse on Method*, 73.
2 Ibid., 117.
3 Maehle and Tröhler. "Animal Experimentation from Antiquity to the End of the Eighteenth Century," 26.
4 Guerrini, "The Ethics of Animal Experimentation in Seventeenth-Century England," 395. Guerrini uses data from Thomas Birch's eighteenth-century *History of the Royal Society of London.*
5 Frank, *Harvey and the Oxford Physiologists*, 129–135.
6 Boyle, *Works,* vol. 2, 17. Cited in Guerrini, 396.
7 Hooke, 53. Cited in Jardine, *Ingenious Pursuits*, 116. Hooke's resolve did not last. He gave in to repeated pleas to perform the experiment once more in October 1667.
8 For an excellent overview of approaches to the soul in the Renaissance and seventeenth century, see Garber, "Soul & Mind," esp. 759–764.
9 Descartes, *Traité de l'homme.* AT XI 174.
10 Wood, *The Life and Times of Anthony Wood,* vol. 2, 12.
11 Frank, *Harvey and the Oxford Physiologists*, 183.
12 Ibid., 182.
13 Lower to Boyle, 18 January 1662.
14 Lower to Boyle, 24 June 1664. See also Frank, *Harvey and the Oxford Physiologists*, 174–175.
15 Lower to Boyle, 8 June 1664. See also Frank, *Harvey and the Oxford Physiologists*, 174–175.

Chapter 4: PLAGUE AND FIRE

1 Yeomans, *Comets*, 69; see also Schechner, *Comets*, 17–24.
2 Aristotle, *Meterologica*, 7.
3 Manilius, *Astronomica* 1, 893–895.
4 Schechner, *Comets*, 62.
5 Moote and Moote, *The Great Plague*, 20.
6 Evelyn, *Diary*, vi. 3, 638. See also Cockayne, *Hubbub*, 107–130.

7 Evelyn, *Fumifugium*, "To the Readers," 5, 10. See also Cockayne, *Hubbub*, 181–205.

8 Moote and Moote, *The Great Plague*, 11.

9 Thomas Vincent, *God's Terrible Voice in the City (1667)*. See also Cockayne, *Hubbub*, 157.

10 Pepys, *Diary*, 13 August 1665, 14 September 1665.

11 Moote and Moote, *The Great Plague*, 177.

12 Ibid.

13 Cockayne, *Hubbub*, 214. In his lively treatise *Fumifugium: or the inconvenience of the aer and smoak of London dissipated* (1661), John Evelyn makes a passionate case against "horrid stinks, niderous and unwholesome smells which proceed from the Tallow, and corrupted Blood" of candlemakers and butchers.

14 "The Method Observed in Transfusing the Bloud out of One Animal into Another," *Philosophical Transactions* 20 (December 17, 1666): 353–358, and Lower, *Tractatus de corde*, 172–176.

15 "The Method Observed in Transfusing the Bloud out of One Animal into Another." *Philosophical Transactions* 20 (December 17, 1666): 353.

16 Boyle to Lower, 26 June 1666; Lower to Boyle, 3 September 1666; Lower, *Tractatus de corde*, 177–179. See also Frank, *Harvey and the Oxford Physiologists*, 177–178.

17 "The Method Observed in Transfusing the Bloud out of One Animal into Another," *Philosophical Transactions* 20 (December 17, 1666): 356.

18 Tinniswood, *By Permission of Heaven*, 46.

19 Reddaway, *Rebuilding London After the Great Fire*, 23.

20 Debris may still have been smoking as late as March 1667. See Dolan, "Ashes and the 'Archive,' " 382.

21 Jardine, *Robert Hooke*, 135; Hall places the number around eighty thousand (*Henry Oldenburg*, 111).

22 Taswell, *Autobiography*, 11.

23 See Linebaugh, "The Tyburn Riot Against the Surgeons"; Porter, *The Great Fire of London*, 87–90; and Tinnisword, *By Permission of Heaven*, 163–168.

24 *Calendar of State Papers, Domestic Series, of the Reign of Charles II,* vol. 6

(1666–1667), 175. Cited in Dolan, "Ashes and the 'Archive,'" 392. See also Bedloe, *Narrative and Impartial Discovery*, 1–19.

25 Oldenburg to Boyle, 10 September 1666.

26 Jardine, *On a Grander Scale*, 239–247.

27 See ibid., 263.

28 Of course these early automata were not intended to actually create artificial life through machinery as much as they were meant to represent visually the mechanistic functions underlying biological processes.

29 "Machine surprenante de l'homme artificiel du sieur Reyselius," *Journal des sçavans*, December 20, 1677: 252.

30 "Statua humana circulatoria," *Journal des sçavans*, November 21, 1683: 317.

31 Hollis, *London Rising*, 142.

Chapter 5: PHILOSOPHICAL TRANSACTIONS

1 After the fire the Royal Society had been obliged to move to a new home at Arundel after the city had commandeered Gresham College for its own postfire administrative needs. It was at Arundel Place that this experiment was performed.

2 D. C. Martin, "Former Homes of the Royal Society," 13.

3 *Philosophical Transactions*, November 19, 1666: 352.

4 Hall, "Oldenburg and the Art of Scientific Communication," 288.

5 Hall, *Henry Oldenburg*, 80.

6 Brown, *Scientific Organizations*, 155; Justel to Oldenburg, 26 May 1666.

7 Language barriers frequently posed challenges for scientific communication between France and England. While periodicals such as the *Philosophical Transactions* and the *Journal des sçavans* were published in the vernacular, many natural philosophers continued to write in Latin as a way to ensure the broadest and most scholarly reach for their works. Still, Latin itself had not proved helpful in facilitating face-to-face communication. As the ever-frank Sorbière remarked in 1663, the English "speak Latin with such an Accent and Way of Pronunciation that they are as hard to be understood, as if they spoke their own language"—a

critique that was shared, no doubt, by the English themselves about their French interlocutors. Sorbière, *A Voyage to England* (1709), 38.

8 Denis to Oldenburg, 22 June 1668.

9 Modern-day Paris has only two islands: the Île de la Cité and the Île Saint-Louis. Earlier islands have either disappeared or were annexed by these two main islands.

10 Brugmans, *Séjour de Christian Huygens*, 26.

11 Denis was married on October 3, 1666. We know very little about his wife. See Peuméry, *Jean Denis*, 9.

12 Denis, *Relation curieuse d'une fontaine*, 349.

13 Brockliss and Jones, *Medical World*, 86–90.

14 Cited in Jones, "Medicalisation," 61.

15 Brockliss and Jones, *Medical World*, 142.

16 Cited in ibid., 140.

17 Ibid., 141.

18 In 1663 Gui-Crescent Fagon made still another plea for circulation. In his thesis *An sanguine impulsum cor salit? Aff.* (Does the heart beat from a pulsation of the blood? Affirmative), in an original assertion, he argued that fetuses first developed hearts rather than livers, as had long been believed. The faculty conceded this point, but full acceptance of circulation was far from forthcoming. See Roger, *The Life Sciences*, 30–31.

19 "Extrait du Journal d'Angleterre contentant la manière passer le sang d'un animal dans un autre," *Journal des sçavans*, January 31, 1667: 31–36.

20 Denis, *Journal des sçavans*, March 9, 1667: 69–72.

21 Ibid., 71.

22 Ibid.

23 Ibid., 72.

24 Oldenburg to Boyle, 10 June 1663.

Chapter 6: NOBLE AMBITIONS

1 See Schechner, "Material Culture," 189–222. Nicolas de Blegny's guide to Parisian luxury goods emphasizes the Quai de l'Horloge as a premier destination for scientific instruments and also confirms presence of the

fish vendors on the riverbanks below (*Livre commode des addresses de Paris,* 1671). Jean-Dominique Augarde offers more details, including addresses, of the shops selling scientific instruments on the quai in the late seventeenth century ("La Fabrication des instruments scientifiques," 60–61).

2 Brown, *Scientific Organizations*, chap. 4, and Turner, *Early Scientific Instruments*, 183.

3 Blegny, *Livre Commode*, 149, 295.

4 Turner, *Early Scientific Instruments*, 183.

5 Schechner, "Material Culture," 209.

6 Ibid., 211.

7 Ibid., 212.

8 Ibid., 210–214; Turner, *Scientific Instruments*, 170, 175.

9 Brown, *Scientific Organizations*, 89.

10 DeJean, *Essence of Style*, 124–126.

11 Now Canada.

12 Chocolate was not only for dessert; it also had medicinal purposes. Physicians to these noble families had begun prescribing regular doses of cocoa syrup for coughs, sore throats, and heartburn—which made chocolate an ideal after-dinner digestif. See Blegny, *Le Bon usage du thé, du caffé.*

13 DeJean, *Essence of Style*, 135.

14 See Brown, *Scientific Organizations*, and Delorme, "Un Cartésien ami."

15 Lennon, *The Battle of the Gods and Giants*, 7.

16 Ibid., 9.

17 Brown, *Scientific Organizations*, 68.

18 Taton, "Huygens et l'Académie royale des sciences," 57.

19 Van Helden, "Saturn and His Anses," 107.

20 Andriesse, *Huygens*, 122–123.

21 Van Helden, "Saturn and His Anses," 111.

22 The original anagram can be found in Huygens, *Oeuvres completes,* vol. 15, 177. See also Howard, "Rings and Anagrams: Huygens System of Saturn," 485.

23 Chapelain, letter no. 304, June 23, 1565.

24 Van Helden, "Telescope in the Seventeenth Century," 43; Westfall, "Henri-Louis de Montmor."

Chapter 7: "HOW HIGH WILL HE NOT CLIMB?"

1 Madame de La Fayette, *Princesse de Clèves*, 41.

2 Astier, "Louis XIV," 74.

3 Motteville, *Memoirs*, vol. 2, 47.

4 Ranum, *Paris in the Age of Absolutism*, 276.

5 Kettering, *French Society*, 193–194.

6 Ranum, *Paris in the Age of Absolutism*, 331.

7 Ibid., 332; La Fontaine, "À M. de Maucroix. Relation d'une fête donnée à Vaux. 22 août 1661"; La Fontaine, "À M. de Maucroix. Ce samedi matin, septembre 1661," *Oeuvres diverses de la fontaine*, 180.

8 La Fontaine, "À M. de Maucroix," 22 August 1661.

9 *Le Nord* was an appellation that Madame de Sévigné gave Colbert following her failed attempts to engage him in her large social circles.

10 Trout, *Jean-Baptiste Colbert*, 40.

11 For details of Louis' weeks-long coordination of Fouquet's eventual arrest, see Dessert, *Fouquet*, 231–262.

12 Madame de Sévigné, "Lettre à Pomponne," 24 November 1664.

13 Montclos, *Vaux Le Vicomte*, 147, and Voltaire, *Le Siècle de Louis XIV*, chap. 25, 277–279.

14 Brown, *Scientific Organizations*, 77, 86.

15 "Discours sceptique en faveur des bêtes et du gouvernement despotique," *Les Mémoires de l'Abbe de Marolles*, n.p. See Morize, "Samuel Sorbière (1610–1670)," 241.

16 Hobbes, *Leviathan*, 186.

17 Without strict regulations, Sorbière wrote to Hobbes, "I fear that what happens to our Montmorian Academy . . . will come to confirm your political theories, and that the less we achieve in natural sciences, the more we prove, by actual practice, the complete truth of your most subtle elements of political philosophy." Sorbière to Hobbes, 12 May 1661, Hobbes, *Correspondence*, vol. 2, 896. See also Adkins, "The Montmor Discourse."

18 Sorbière's bylaws for the Montmor Academy are reprinted in Bigourdan, *Premières sociétés savantes*, 13–14.

19 Sorbière's speech of 1663 to the Montmor Academy is reprinted in ibid., 14–20.
20 Ibid., 15.
21 Brown, *Scientific Organizations*, 133.
22 Madame de Sévigné, "Lettre à Pomponne," 18 December 1664.
23 Collas, *Jean Chapelain*, 361–369.
24 Collas, *Jean Chapelain*, 383–388.
25 Sturdy, *Science and Social Status*, 74–75.
26 Hahn, "Changing Patterns," 407.
27 Brown, *Scientific Organizations*, 133; Huygens, *Oeuvres complètes*, vol. 5, 70.

Chapter 8: THE KING'S LIBRARY

1 Senior, "The Menagerie," 210.
2 Kalof, *Looking at Animals*, 122.
3 Lister, cited in Stroup, *A Company of Scientists*, 41.
4 Brockliss, "Medical Teaching," 247, and Brockliss, "University of Paris," 230.
5 Perrault, *Memoirs of My Life*, 20–30.
6 All details of the experiments and outcomes are drawn from the Académie des Sciences archive manuscript coauthored by Perrault, Gayant, and Auzout (22 January 1667); Perrault, "De la transfusion du sang"; and "An Account of More Tryals of Transfusion," *Philosophical Transactions*, October 21, 1667.
7 Perrault, "De la transfusion du sang," 428, 429–430, 437.
8 Ibid., 425–426.
9 Picon, *Claude Perrault*, 223–230; *Catholic Encyclopedia*, vol. 1, ed. Charles Herbermann, s.v. "Perrault."

Chapter 9: THE PHILOSOPHER'S STONE

1 Claude Perrault, "De la transfusion du sang," 423.
2 Ovid, *Metamorphoses*, book 7.
3 The overlap of the two fields would continue well into the eighteenth

century, when chemistry would begin to shift more clearly toward medical chemistry or what we now call pharmacy—which helps to explain why, in England, the popular word for pharmacist is still chemist. Newman and Principe, "Alchemy vs. Chemistry," 38, 39. See also Principe, *The Aspiring Adept*, 107–111.

4 Hall, *Henry Oldenburg*, 67.

5 Boyle, "Incalescence," 528, 529.

6 "Tryals Proposed by Mr. Boyle to Dr. Lower . . . for the Improvement of Transfusing Blood Out of One Live Animal into Another," *Philosophical Transactions* 22 (February 11, 1666): 385–388.

7 Ibid., 386–387.

8 Odoric, *Travels*, 114.

9 See Bondeson, *Two-Headed Boy*, 2. See also Wiesner, *Marvelous Hairy Girls*.

10 Bondeson, *Two Headed-Boy*, 99.

11 Ibid., 37, 116.

12 The medieval travel writer Bartholomaeus Anglicus hedged all bets in his *On the Nature of Things* (c. 1230) and placed them under the rubrics of both "man" and "animal." His volume drew so much interest that the Sorbonne was obliged to chain its manuscript copy of Bartholomaeus's text to prevent it from being stolen by readers. Ramey, "Monstrous Alterity," 86.

13 Cited in Riesman, "Bourdelot," 191.

14 Pepys, *Diary*, 14 November 1666.

15 Porter, *Greatest Benefit*, 201.

16 Ackerknecht, *Short History of Medicine*, 56.

17 Unlike in France, where iatrochemistry was still controversial, prominent natural philosophers like Boyle and Willis had come out strongly in favor of it. Hall, "English Medicine in the Royal Society's Correspondence," 116.

18 Debus, *French Paracelsans*, 21–23.

19 Cited in ibid., 99.

20 Perrault, "De la transfusion du sang," 409.

Chapter 10: THE BLOOD OF A BEAST

1 Sorbière discourse, 1663. Reprinted in Bigourdan, *Premières sociètes savantes*, 18.

2 "A Letter Concerning a New Way of Curing Sundry Diseases by Transfusion of Blood, Written to Monsieur de Montmor," *Philosophical Transactions*, June 25, 1667.

3 *Journal des sçavans*, April 8, 1667, 96. Reprinted in *Philosophical Transactions*.

4 Oldenburg to Stanislas Lubienietzki, 3 January 1667.

5 "An Account of an Easier and Safer Way of Transfusing Blood," *Philosophical Transactions*, May 6, 1667: 451; "An Account of Another Experience of Transfusion, viz. of Bleeding a Mangy into a Sound Dog," *Philosophical Transactions,* May 6, 1667: 451–452.

6 "An Account of an Easier and Safer Way of Transfusing Blood," *Philosophical Transactions,* April 18, 1667: 450.

7 The Italians received news of Denis' transfusions via the *Philosophical Transactions* and the *Journal des sçavans.* Further demonstrating Italian interest in the procedure, the majority of Denis' reports were published, in translation, in 1668 by Emilio Maria Manolessi in his *Relazione dell'Esperienze Fatte in Inghilterra, Francia, ed Italia Intorno alla celebre e famosa trasfusione del sangue.*

8 "An Extract out of the Italian Giornale de Letterati, about Two Considerable Experiments of the Transfusion of the Blood." *Philosophical Transactions,* January 1, 1668, 840–842.

9 "Esperienze Fatte in Roma per la Trasfusione del Sangue," in Manolessi, *Esperienze*, 36.

10 See Manfredi's dedication to Marie Mancini, who after marriage became Marie Colonna, in *De Nova et inaudia medico-chyrurgica operatione sanguinem transfudente de individuo ad individuum*, 1668.

11 "Concerning a New Way of Curing Sundry Diseases by Transfusion of Bloud," *Philosophical Transactions,* June 25, 1667: 158–159.

12 Ibid.

13 Woolley, *The Queen-Like Closet*, 11; W. M., *The Queen's Closet Opened*, 7–8.

14 Levy-Valensi, *La Médecine*,123–125.

Chapter 11: THE TOWER OF LONDON

1 A few writers indicate that he was a *crocheteur*, a hook maker. The most reliable accounts indicate, however, that he was a butcher. References to a hook maker are not necessary inconsistent butcher references; hooks are used to hang meat for smoking, storage, and preparation, suggesting the man's butcher connections.

2 Hussey, *Paris: The Secret History*, 162–163.

3 Zeller, "French Diplomacy and Foreign Policy in Their European Setting," 202.

4 Oldenburg to Boyle, 16 December 1667.

5 Hall, *Henry Oldenburg*, 85–86.

6 Browne, *Getting the Message*, 18. Sometimes foreign correspondents would "frank" the fees; that is, pay the postage up until their country's border. This would, however, not been entirely helpful for Oldenburg because he would be expected to prepay future outbound letters to Dover in kind. Hall, *Henry Oldenburg*, 83.

7 The original French version of Denis' letter is dated June 25. Yet Hall and Hall (*Priority Disputes*) speculate, as do I, that Oldenburg already had a copy of it, perhaps in manuscript, as early as June 20. In his letter to Boyle dated 25 September, 1667, Oldenburg explained that "I had the French Original before anybody had it in England, which was the same day of my confinement." Oldenburg was arrested on June 20, presumably the same day he received Denis' letter.

8 "Concerning a New Way of Curing Sundry Diseases by Transfusion of Blood," *Philosophical Transactions*, June 25, 1667: 489–490. Denis' emphasis.

9 Wallis to Oldenburg, 21 March 1667.

10 Pepys, *Diary*, 16 June 1667.

11 Marshall, *Intelligence and Espionage*, 78.

12 Quoted in Dickinson, *Sir Samuel Morland*, 96. Morland also devised "a most dextrous and expeditious way of copying out any sheet of paper close written on both sides in little more than a minutes time" for the task. Marshall, *Intelligence and Espionage*, 86.

13 Scheider, *Culture of Epistolarity*, 83.

14 Jardine, *Ingenious Pursuits,* 323. For urine as invisible ink, see Boyle, *Memoirs for the Natural History of Humane Blood*, 256. Boyle also later tried using blood serum as ink. For techniques to reveal obscured letters, see Schneider, *The Culture of Epistolarity*, 83.

15 Marshall, *Intelligence and Espionage*, 80.

16 Hall, *Henry Oldenburg*, 113, 117.

17 Pepys, *Diary*, 25 June 1667.

18 On the timeline of Oldenburg's receipt of Denis' letter, see note 7 above.

19 See Webster, "The Origins of Blood Transfusion: A Reassessment," 387–391.

20 Clarke to Oldenburg, April/May 1668, *Philosophical Transactions*, May 18, 1668: 672–682.

21 July 4, 1667. See McKie, 32–33.

22 Oldenburg to Seth Ward (for the Bishop of Exeter to be forwarded to the Bishop of Salisbury, via an unknown scribe), 15 July 1667.

23 Hall, *Henry Oldenburg*, 117.

24 Ibid., 120, 145–149.

25 Oldenburg to Boyle, 3 September 1667.

26 Oldenburg described Lower's visit in his letter to Boyle dated 24 September 1667.

27 "An Advertisement Concerning the Invention of the Transfusion of Blood," *Philosophical Transactions*, July, August, September 1667: 489–490.

28 "An Account of More Tryals of Transfusion," *Philosophical Transactions,* December 9, 1667: 517–519.

29 Justel to Oldenburg, 6 November 1667.

Chapter 12: BEDLAM

1 "An Account of an Easier and Safer Way of Transfusing Blood out of One Animal to Another," *Philosophical Transactions,* May 6, 1667.

2 Birch, *History of the Royal Society*, vol. 2, 201. All references to the Royal Society's discussion and plans to begin human transfusion are from Birch's *History*, a contemporary record of the society's activities.

3 The diarist John Evelyn confirmed living conditions in Bedlam after a visit to the hospital on April 21, 1656. For a detailed study of madness in seventeenth-century England, see McDonald, *Mystical Bedlam.*

4 Boyle, *Some Considerations,* vol. 2, 58–59, cited in Lawrence and Shapin, *Science Incarnate*, 97.

5 There was no set of normative ethics that distinguished medical research from medical treatment in the early modern era. For a detailed study of the ethics of human experimentation, see Lederer, *Subjected to Science.*

6 Macdonald, *Mystical Bedlam*, 190–191.

7 King to Boyle, 25 November 1667. King's account, "Of the Experiment of Transfusion Practiced upon a Man in London," was also published in the *Philosophical Transactions.*

8 King to Boyle, 25 November 1667. My description of the procedure is based on King's detailed account.

9 Ibid.

10 Oldenburg to Boyle, 25 November 1667.

11 Ibid., 25 November 1667.

12 Pepys, *Diary*, 30 November 1667.

13 Oldenburg to Sluse, 25 November 1667.

14 Birch, *History of the Royal Society*, vol 2, December 12, 1667.

15 Oldenburg to Boyle, 17 December 1667.

16 Printed in Stubbe, *A Specimen*, 179.

17 Nicolson, *Pepys' Diary and the New Science.*

18 John Skippon to John Ray, January 24, 1668. Cited in Nicolson, *Pepys' Diary and the New Science*, 169.

19 Shadwell, *The Virtuoso*, act 2, ll. 206–211.

20 Académie des Sciences archive manuscript, Perrault, Gayant, Auzout, 22 January 1667.

21 "Cure of an Inveterate Phrensy by the Transfusion of Bloud," 618.

22 Twigg, *Bathing*, 25–28.

Chapter 13: MONSTERS AND MARVELS

1 All references to Martinière's adventures as a surgeon, pirate, and slave are from his autobiography, *Heureux esclave.*
2 Martinière, *Heureux esclave*, 13.
3 Dannenfeldt, "Egyptian Mumia," 173.
4 Martinière, *Heureux esclave*, 119.
5 Ibid., 204.
6 Martinière, *Les sentiments d'un vray médicin*, 5.
7 Ibid., 4.
8 Ibid., 5.
9 Martinière, *L'Ombre d'Apollon*, 3. The account is dated September 15, 1667.
10 Ibid., 3–4.
11 "A Letter Concerning a New Way of Curing Sundry Diseases by Transfusion of Bloud, Written to Monsieur de Montmor"; "Lettre de G. Lamy à M. Moreau Docteur en Médecine de la Faculté de Paris contre les pretendues utilitez de la Transfusion"; "Lettre de G. Gadroys à M. l'Abbé Bourdelot Docteur de Médecine de la Faculté de Paris, pour servir de Réponse à la lettre écrite par M. Lamy contre la Transfusion."
12 Lamy, "Lettre écrite à Moreau."
13 Ibid.
14 Lamy, *Discours anatomiques*, cited in Kors, "Monsters and the Problem of Naturalism," 36.
15 Lamy, *Discours anatomiques*, 22–23.
16 See Debus, *French Paracelsans*, 84–95.
17 Brygoo, "Les médecins de Montpellier," 7. See also Jones, "Medicalisation."
18 Martinière, *L'Ombre d'Apollon*, 16.

Chapter 14: THE WIDOW

1 Hussey, *Paris: The Secret History*, 165–166.
2 "An Extract of a Printed Letter, Addressed to the Publisher, by M. Jean

Denis. . . . Touching the Differences Risen About the Transfusion of Bloud," *Philosophical Transactions*, 1668, vol. 3: 710–715. Descriptions of the circumstances of Mauroy's death have been distilled from this and other contemporary reports, including extant court records.

3 Peuméry, *Jean-Baptiste Denis*, 127–137.

4 "An Extract of a Printed Letter, Addressed to the Publisher, by M. Jean Denis. . . . Touching the Differences Risen About the Transfusion of Bloud," *Philosophical Transactions*, 1668, vol. 3: 711.

5 Ibid.

6 Justel to Oldenburg, 15 February 1668.

7 Justel to Oldenburg, 25 February 1668.

8 Justel to Oldenburg, 21 February 1668.

Chapter 15: THE AFFAIR OF THE POISONS

1 Denis, "Extract of a Letter Touching a Late Cure." An "Extrait des registres du greffe criminal du Chastelet de Paris, du mardy 17 avril 1668" was included in Denis' letter and published in *Philosophical Transactions* 3 (1668), 710–715, and reprinted as well in Martinière, *Remonstrances Charitables.* While the original decree is no longer extant, the "extraits" published by Denis and Martinière and elsewhere are identical and reasonably reliable, particularly given that there is no debate between the two men about the faithfulness of the document in regard to the official proceedings at the Grand Châtelet.

2 "An Extract of a Printed Letter, Addressed to the Publisher, by M. Jean Denis. . . . Touching the Differences Risen About the Transfusion of Bloud," *Philosophical Transactions*, 1668, vol. 3: 711.

3 The corpses were laid out, usually uncovered, in one of the Châtelet's many open courtyards until they could be collected, washed, and transferred (by the nuns of nearby Saint-Opportune hospital) to the pit graves in the cemetery that was steps away from the Grand Châtelet. Chardans, *Le Châtelet*, 36. See also Andrews, *Law, Magistracy*, 14.

4 Chardans, *Le Châtelet*, 42.

5 Saint-Germain, *La Reynie*, 36.

6 Twice a week the *Lieutenant Criminel* (criminal lieutenant) presided over the presentation of the facts in minor criminal cases, which were punishable by the payment of damages and court costs. Mousnier, *Institutions of France under the Absolute Monarchy*, 319.

7 "Extrait des registres du greffe criminal du Chastelet de Paris, du mardy 17 avril 1668." See also Denis, "Extract of a Letter Touching a Late Cure."

8 "Extrait des registres du greffe criminal du Chastelet de Paris, du mardy 17 avril 1668."

9 Ibid.

10 Ibid.

11 Cited in Bluche, *Louis XIV*, 268.

12 Mollenauer, *Strange Revelations*, 72.

13 "Extrait des registres du greffe criminal du Chastelet de Paris, du mardy 17 avril 1668."

14 Ibid.

15 Ibid.

16 Denis to Oldenburg, May 5, 1668.

17 See Hamscher, *The Parlement of Paris*, 98–107; Hamscher, *Conseil Privé*, esp. chap. 2; and Andrews, *Law, Magistracy*, chap. 2, for these and other details regarding court structure and protocol at parliament.

18 Confirmed by Lamy, "Lettre escrite a Monsieur Moreau," February 16, 1668, 1.

19 The attendees are confirmed in "A Letter Written by an Intelligent and Worthy Englishman from Paris," *Philosophical Transactions*, December 13, 1669, 1075. They are consistent with the traditional composition of the Grand'Chambre. See Andrews, *Law, Magistracy*, 88.

20 The last *lit de justice* was held on February 24, 1673. It deprived the parliament of Paris of the right to question royal edicts before officially registering them.

21 "A Letter Written by an Intelligent and Worthy Englishman from Paris," 1075. My colleague Albert Hamscher, an internationally known specialist on the seventeenth-century parliament, confirmed that

records not only for this parliamentary case but thousands of others during the same time period were destroyed or lost at some point over the hundred years or so that followed.

Chapter 16: CHIMERAS

1 Oldenburg confirmed this to Denis in a letter, now lost, dated 29 April 1668.
2 Lower, *De corde*. Cited in Franke, *Oxford Physiologists*, 214.
3 Martinière, *Sur l'Ombre de Phaeton*, 4.
4 Lamy, "Lettre escrite à Monsieur Moreau," 16 February 1668, 7–8.
5 Ibid., 10.
6 Ibid., 11.
7 Martinière, *Sur l'Ombre de Phaeton,* Preface, 1.
8 See Thomson, "Guillaume Lamy et l'âme materielle," 64–70.
9 Martinière, *Sur l'Ombre de Phaeton*, 2–3.
10 Ibid., 3.
11 Martinière, *Chymique ingénue*, 74.
12 Martinière, *Rencontres de Minerve*, 4.
13 Martinière, *Médée resuscitée*, 4.
14 Martinière, *Remonstrances charitables*, 8-12.
15 See Mayor, *The Poison King*, 220, 237–238.
16 Martinière, Letter to Molé, 2.
17 Martinère, *Les sentimens d'un vray médecin*. 1.
18 Martinière, *Rencontres de Minerve*, 2.
19 Martinière, *Les sentimens d'un vray médecin*, 6.
20 Lamy, "Lettre escrite à Monsieur Moreau," 10: "Il [Denis] s'est creû vivement offencé de ce que j'ai tâché quoiqu'innocemment d'étouffer dès le berceau les esperences (He [Denis] believed himself to be so fully insulted by my efforts, however innocent, to suffocate in the cradle [his] hopes"); Martinière, *Remonstrances charitables*, 1: "Pour la peine que je prends à tâcher d'étouffer dans le berceau ce Monstre transfusionnaire" (For the trouble that I take to suffocate in the cradle this transfusionist Monster").

21 Martinière, Letter to Molé, 1.
22 Martinière, Letter to Colbert, 1.
23 Ibid., 3.
24 Ibid.
25 Ibid.
26 Loux, *Martinière*, 15.
27 Chapelain to Régnier de Graff, 28 August 1671. It was the same poet Chapelain who helped orchestrate the demise of Montmor's private academy, providing Colbert with a list of men he might poach.
28 Chapelain to Régnier de Graff, 28 August 1671.
29 Foiret, "L'Hôtel de Montmor," 320.
30 Denis, *Discours sur l'astrologie judiciaire,* 1, 36.
31 Denis, Conference of April 30, 1673, reported in *Philosophical Transactions* on May 30, 1673; "Extract of a Letter, Written to the Publisher by M. Denys from Paris; Giving Notice of an Admirable Liquor, Instantly Stopping the Blood of Arteries Prickt or Cut, Without Any Suppuration, or Without Leaving Any Scar or Cicatrice," *Philosophical Transactions*, 1673, vol. 8: 6039, "Experiments Made at London Concerning the Liquor Sent out of France, Which is There Famous for Staunching of the Blood Arteries as Well as Veins," *Philosophical Transactions*, 1673, vol. 8: 6052–6059.
32 Blundell, "On the Transfusion of Blood."
33 Blundell, *The Principles and Practice of Obstetricy*, 247, 337, 580, 838. Blundell acknowledged that he was also inspired by the work of John Henry Leacock, who was experimenting with animal-to-animal transfusions at the same time.
34 Blundell, "On the Transfusion of Blood," 60.
35 Ibid., 75. The idea that animal blood could be transfused to humans came definitively to an end in the mid-1870s. E. Ponfick, a pathologist, confirmed that mixing blood of two different species led to lysis, or cell disintegration, resulting in a life-threatening reaction in the recipient. L. Landois reviewed all known transfusions in humans—478 in all—from the earliest beginnings in the cases of Denis and Lower. One hundred and twenty twenty-nine of the donors had been animals,

the remainder were human donors. He calculated that one-third of the patients receiving animal blood survived. The results were more promising for human-to-human transfusions, where more than one-half survived. Diamond, "A History of Blood Transfusion," 672–673. As Diamond notes, there were advocates for animal donors in human transfusions as late as 1928.

36 Starr, *Blood*, 37.

37 Schneider, "Transfusion in Peace and War," 113.

38 Other attempts included the use of leech saliva, which contains the anticoagulant hirudin, or allowing the blood to clot and then scooping out the clots from the blood before administering it to the recipient. Diamond, "A History of Blood Transfusion," 671–672. And in the late 1860s, the Englishman John Braxton Hicks—who is better known for his work on obstetric contractions—added phospate of soda, a common ingredient in early fountain drinks, to donor blood. While it did reduce clotting, it proved fatal to the recipient in every case.

39 Schneider, "Blood Transfusion Between the Wars," 212.

40 Schmidt, "American Association of Blood Banks," 93.

41 Hillyer, "The Blood Donor," 25, and http://www.redcrossblood.org/learn-about-blood/blood-facts-and-statistics (accessed March 6, 2010).

Epilogue

1 Cited in Bonnicksen, *Chimeras, Hybrids, and Interspecies Research*, 5.

2 Greely, "Thinking about the Human Mouse."

3 Since the 1980s, the hamster oocyte penetration test has been used to evaluate the viability of human sperm. The sperm is introduced to hamster eggs kept in culture. If penetration occurs, a one-celled "humster" zygote results and is destroyed before it divides any further. The first, and only, "geep" was created in 1984. Bonnicksen, *Chimeras, Hybrids, and Interspecies Research*, 10, 50, 55.

4 Bonnicksen, *Chimeras, Hybrids, and Interspecies Research,* esp. chap. 1. For a productive analysis of the relationship between scientific and literary chimeras, see also Clayton, "Victorian Chimeras."

5 Karpowitz et al., "Developing Human-Nonhuman Chimeras"; Greene et al., "Moral Issues of Human-Non-Human Primate Neural Grafting"; Greely, "Thinking about the Human Neuron Mouse"; Bonnicksen, *Chimeras, Hybrids, and Interspecies Research.*

Bibliography

PRIMARY SOURCES

Académie des Sciences. "Observations anatomiques faites en Assemblée qui se tient dans le logis ou à la bibliothèque du Roy par ordre de sa majesté." Ms. 22, January–March 1667.

Aristotle. *Meteorologica.* Translated by H. D. P. Lee. Vol. 1. Cambridge, MA: Loeb Classical Library/Harvard University Press, 1978.

Aubrey, John. *Brief Lives.* Edited by Oliver Lawson Dick. Ann Arbor: University of Michigan Press, 1957.

Bacon, Francis. *The Works of Francis Bacon.* 17 vols. Edited by Basil Montagu. London, 1830.

Basril, Louis. "Réflexions de Louis de Basril, advocate en Parlement, sur les disputes qui se font á l'occasion de la transfusion." n.p., n.d. Bibliothèque interuniversitaire de Médecine, 160725; Bibliothèque de l'Arsenal, 4-S-2229(12).

Bedloe, William. *A Narrative and Impartial Discovery of the Horrid Popish Plot, Carried On for the Burning and Destroying of the Cities of London and Westminster, with their Suburbs.* London, 1679.

Bie, Jacques de. *La France métallique contenant les actions célèbres tant publiques que privées des rois et des reines.* Paris, 1636.

Birch, Thomas. *History of the Royal Society of London for Improving of Natural Knowledge.* 4 vols. London, 1756–1757.

Blegny, Nicolas de. *Le Bon usage du thé, du caffé et du chocolat pour la préservation & pour la guérison des maladies.* Paris, 1687.

———. *Livre commode des addresses de Paris.* Paris, 1671.

Blundell, James. "Experiments on the Transfusion of Blood by the Syringe." *Medico-Chirurgical Transactions* 9 (1818): 56–92.

———. "Observations on the Transfusion of Blood by Dr. Blundell. With a Description of His Gravitator." *Lancet* 2 (1828–1829): 321–324.

———. "Some Account of a Case of Obstinate Vomiting in which an Attempt was Made to Prolong Life by the Injection of Blood into the Veins." *Proceedings of the Royal Society of Medicine* 10 (1819): 296–311.

———. "Successful Case of Transfusion." *Lancet* 1 (1828): 431–432.

——— *The Principles and Practice of Obstetricy, as at Present Taught by James Blundell.* London, 1834.

Boileau-Despréaux, Nicolas. *Oeuvres.* Paris, 1888.

Bovell, J. "On the Transfusion of Milk, as Practiced in Cholera, at the Cholera Sheds." *Canadian Journal* 3 (1855): 188–192.

Boyle, Robert. *A Disquisition about the Final Causes of Natural Things: Wherein it is Inquired, Whether and (If At All) with What Cautions a Naturalist Should Admit Them?* London, 1688.

———. *Correspondence.* Edited by Michael Hunter, Antonio Clericuzio, and Lawrence M. Principe. 6 vols. London: Pickering & Chatto, 2001.

———. *Memoirs for the Natural History of Human Blood, Especially the Spirit of that Liquor.* London, 1683.

———. *Some Considerations Touching the Usefulness of Experimental Natural Philosophy.* Oxford, 1663.

———. *The Skeptical Chemist.* [1661]. London, 1911.

———. *Works.* 7 vols. Edited by Michael Hunter and Edward B. Davis. London: Pickering & Chatto, 1999.

Bull, William T. "On the Intra-Venous Injection of Saline Solutions as a Substitute for Transfusion of Blood." *Medical Record* 25 (1884): 6–8.

Calendar of State Papers, Domestic Series, of the Reign of Charles II. Vol. 6. 1666–1667.

Reprint. London: Longman, Green, Longman & Roberts, 1860–1938.

Catalogue de tous les livres du feu M. Chapelain, Bibliothèque Nationale, Fonds français, Nouv. Acq. No. 318. Edited by Colbert Searles. Palo Alto: Stanford University Press, 1912.

Chéreul, Adolphe. *De l'administration de Louis XIV, d'après les mémoires inédits d'Olivier d'Ormesson*. 1850. Reprint, Geneva: Slatkine, 1974.

Denis, Jean-Baptiste. See also *Journal des sçavans* and *Philosophical Transactions*.

———. *Discours sur l'astrologie judiciaire et sur les horoscopes*. Paris, 1668.

———. *Relation curieuse d'une fontaine découverte en Pologne*. Paris, 1687.

Descartes, René. *The Philosophical Works of Descartes*. 2 vols. Translated by Elizabeth S. Haldane and G. R. T. Ross. London: Cambridge University Press, 1967.

Eloy, Nicolas F. J. *Dictionnaire historique de la médecine*. Mons, 1778.

Evelyn, John. *The Diary of John Evelyn*. 3 vols. 1906. Reprint, London: Routledge, 1996.

———. *Fumifugium: or the Inconvenience of the Aer and Smoak of London Dissipated*. London, 1661.

Ficino, Marsilio. *De vita libri tres (Three Books on Life)*. Translated by Carol V. Kaske and John R. Clark. 1489. Reprint, Binghamton, NY: Medieval & Renaissance Text and Studies, 1989.

Folli, Francesco. *Stadera medica, nella quale oltra la medecine infusoria ed altre novita, si bilanciano le ragioni favorevoli e le contrarie alla transfusione del sangue*. Florence, 1680.

Galen. *Claudii Galeni opera omnia*. Edited by C. G. Kuhn. 1453–. Reprint, Hildesheim: G. Olms, 1964.

Graverol, François. *Sorberiana, ou les pensées de M. de Sorbière*. 2nd. ed. Paris, 1695.

Guyre, Gaspard de. "Lettre écrite à Monsieur l'Abbé Bourdelot, Docteur en Médecine de la Faculté de Paris, Premier Médicin de la Reine Christine de Suède . . . sur la transfusion du sang, contenant des raisons & des experiences pour & contre." Paris, 1667.

Harvey, William. *De motu cordis and De circulatione sanguinis*. Translated by Geoffrey Keynes. New York: Dover Publications, 1995.

———. *Works*. Translated by Robert Willis. London: Sydenham Society, 1847.

Henry, David. *An Historical Account of the Curiosities of London and Westminster, in Three Parts. Part I. The Tower of London.* London, 1767.

Héroard, Jean. *Journal de Jean Héroard.* Edited by Madeleine Foisil. Paris: Fayard, 1989.

Histoire de l'Académie royale des sciences, depuis son établissement au 1666 jusqu'à 1686. Paris, 1733.

Hobbes, Thomas. *The Correspondence.* Edited by Noel Malcolm. 2 vols. New York: Oxford University Press, 1994.

———. *Leviathan.* 1651. Reprint, New York: Penguin Classics, 1968.

Huygens. Christian. *Oeuvres complètes.* The Hague: M. Nijhoff, 1888–1950.

"Injection of Milk into the Veins." *Lancet* 2 (1882): 436.

Journal des sçavans:

"Extrait du *Journal d'Angleterre* contenant la manière de faire passer le sang d'un animal dans un autre." January 31, 1667: 31–36.

"Extrait d'une lettre de M. Denis, Professeur de Philosophie & de Mathematique à M*** touchant la transfusion du sang. De Paris ce 9 mars 1667." March 14, 1667: 69–72.

"Extrait d'une lettre de M. Denis, Professeur de Philosophie & de Mathématique, à M*** touchant la transfusion du sang. Du 2 avril 1667." April 2, 1667: 96.

"Lettre de G. Lamy à M. Moreau Docteur en Médecine la Faculté de Paris, contre les pretenduës utilités de la Transfusion." February 1668: 21–23.

"Lettre de M. Denis, Professeur de Philosophie et de Mathématique, à M. de Montmor, Maistre des Requestes touchant deux experiences de la transfusion faites sur des hommes." June 25, 1667.

"Machine surprenante de l'homme artificiel du sieur Reyselius." December 20, 1677: 252.

"Statua humana circulatoria." November 21, 1683: 317.

Kilduffe, Robert A., and Michael DeBakey. *The Blood Bank and the Technique and Therapeutics of Tranfusions.* St. Louis: C. V. Mosby Company, 1942.

La Fayette, Marie-Madeleine de. *The Princesse of Clèves.* 1678. Translated by Nancy Mitford. New York: New Directions, 1988.

La Fontaine, Jean de. *Oeuvres diverses de La Fontaine*. Paris: Ménard et Desenne fils, 1821.

———. "À M. de Maucroix. Relation d'une fête donnée à Vaux. 22 août, 1661." *Oeuvres diverses de la Fontaine*. Paris: Ménard et Desenne fils, 1821, 172–179.

———. "À M. de Maucroix. Ce samedi matin, septembre 1661." *Oeuvres diverses de la Fontaine*. Paris: Ménard et Desenne fils, 1821, 180.

La Mettrie, Julien Offray de. *Ouvrage de Penelope, ou Machiavel en médecine*. Berlin, 1748–1750.

Le Gallois. *Conversations de l'Académie de Monsieur l'Abbé Bourdelot, contenant diverses recherches, observations, expériences et raisonnements de physique, médecine, chymie et mathématiques*. Paris, 1672.

Libavius, Andreas. *Appendix necessaria syntagmatis arcanorum chymicorum*. Frankfurt, 1615.

Lister, Martin. *A Journey to Paris in the Year 1698*. London, 1699.

Louis, Pierre. *Recherches sur les effets de la saignée dans quelques maladies inflammatoires et sur l'action de l'émétique et des vésicatoires dans la pneumonie*. Paris, 1835.

Lower, Richard. *Tractatus de corde*. London, 1669.

Manfredi, Paolo. *De nova et inaudita medico-chyrurgica operatione sanguinem transfundente de individuo ad individuum; prius in brutis et deinde in homine Romae experta*. Rome, 1668.

———. *Ragguaglio degli'esperimenti fatti sotto la diretitione di Paolo Manfredi, circa la nuova operatione della transfusione del sangue da individuo ad individuo et in bruti et in huomini*. Rome, 1668.

Manilius, Marcus. *Astronomica*. Translated by G. P. Goold, Vol. 1. Cambridge, MA: Loeb Classical Library/Harvard University Press, 1977, 893–895.

Marolles, Michel de. *Les Mémoires de l'Abbé de Marolles*. 1657.

Martinière, Pierre-Martin de la. *Le Chymique ingénu ou l'imposture de la pierre philosophale découverte par le sieur de la Martinière*. n.p., 1665.

———. *Euthryphronis philosophie et medicide nove curandorum morborum ratione per transfusionem sanguinis ratione per transfusionem sanguinis dissertatio ad amicum*. Paris, 1667.

———. *Explication mechanique et physique des fonctions de l'âme sensitive*. Paris, 1677.

———. *L'Heureux esclave ou relation des aventures du Sieur de la Martinière comme il fut pris par les Corsaires de Barbarie & délivré; La manière de combattre sur Mer, de l'Afrique & autres particularitez.* Paris, 1674.

———. *Lettre envoyée à Madame Louyse de Vieupont, veuve du feu Seigneur Louys Doinville, Conseiller du Roy en ses Conseils, Chevalier, Baron de Hoüteville & autres lieux.* Paris, 1667.

———. *Médée ressuscitée, affirmant l'utilité de la transfusion du sang.* Paris, 1668.

——— "À Monseigneur Colbert . . . sur la transfusion du sang." n.d.

———. *L'Ombre d'Apollon découvrant les abus de cette pretenduë manière de guérir les maladies par la transfusion du sang. Ensemble une lettre servant de responce à la première & seconde lettre de Monsieur Denis & Gadroys.* Paris, 1667.

———. *Les Opuscules du Sieur de la Martinière . . . contre les circulateurs & transfuseurs du sang.* Paris, 1668.

———. *Le Prognosticateur charitable traitant des mouvements, natures, regards, conjonctions, dominations et différentes influences tant des planètes que des autres signes célestes.* Paris, 1666.

———. *Rencontres de Minerve, la vertu, honneur et amour, faisans vour l'abus des circulateurs du sang sur le sujet de leur chymere.* Paris, 1668.

———. *Remontrances charitables du Sieur de la Martinière à Monsieur Denis.* Paris, 1668.

———. *Les sentimens d'un vray médecin, faisant voir les inutilitez & cruautez de la transfusion du sang.* Paris, n.d.

———. *Sur l'Ombre de Phaethon, Contre la lettre du Sieur de Montpoly, qu'il a escrite sur le sujet de la transfusion du sang, & contre celle du Sieur Lamy.* Paris, 1667.

———. "A Tres-Haut et Serenissime Cosme Marie. Prince de Toscane." Paris, n.d.

Meurdrac, Marie de. *La Chimie charitable et facile en faveur des dames.* 1666. Reprint, Paris: CNRS, 1999.

Motteville, Françoise Bertaut de. *Mémoires.* Paris: Charpentier, 1869.

Odoric of Pordenone. *The Travels of Friar Odoric.* Translated by Sir Henry Yule. Grand Rapids, MI: Eerdmans, 2002.

Oldenburg, Henry. *The Correspondence of Henry Oldenburg.* Edited by A. Rupert Hall and Marie Boas Hall. 10 vols. Madison: University of Wisconsin Press, 1965–1975.

Ovid. *The Metamorphoses.* Translated by Horace Gregory. New York: New American Library, 1960.

Paracelsus. *Selected Writings.* Edited by Jolanda Jacobi. Translated by Norbert Guterman. New York: Pantheon, 1931.

Paré, Ambrose. *The Workes of that Famous Chirurgion Ambrose Paré.* Translated by T. Johnson. London, 1634.

Pepys, Samuel. *The Diary of Samuel Pepys.* 11 vols. Berkeley: University of California Press, 1970–1983.

Perrault, Charles. *Memoirs of My Life.* Edited and translated by Jeanne Morgan Zarucchi. Columbia: University of Missouri Press, 1989.

Perrault, Claude. *Dossier Claude Perrault. Notes et dessins scientifiques relatifs à son travail à l'Académie des Sciences.* Ms., Académie des Sciences.

———. "Essais de Physique Tom. III de la Mechanique des Animaux par M. Perrault de l'Académie R. des Science D. en M. de la Faculté de Paris." Paris, 1680–1688.

———. "Extrait d'une lettre écrite à Monsieur de la Chambre, qui contient les observations qui ont été faites sur un grand poisson disséqué dans la Bibliothèque du Roy, le vingt-quatrième juin 1667." Paris, 1667.

———. "Observations qui ont été faites sur un Lion disséqué dans la Bibliothèque du Roy, le vingt-huictième Juin 1667, tirées d'une lettre écrite à Monsieur de la Chambre." Paris, 1667.

———. *Ordonnance for the Five Kinds of Columns After the Method of the Ancients.* Translated by Indra Kagis McEwen. Santa Monica, CA: Getty Center for the History of Art and the Humanities, 1993.

———. "Projet des observations anatomiques," ms. 15 Jan. 1667. Pochette des séances, 1667. Archives de l'Académie des Sciences.

———. "Projet pour la botanique," ms. 15 Jan 1667. Pochette des séances 1667. Archives de l'Académie des Sciences.

———. "De la transfusion du sang." *Essais de physique, ou recueil de plusieurs traités touchant les choses naturelles.* Paris, 1680, 404–438.

Philosophical Transactions:

"An Account of an Easier and Safer Way of Transfusing Blood Out of One Animal into Another, viz. by the Veins, Without Opening Any Artery of Either." May 6, 1667: 449–451.

"An Account of Another Experiment of Transfusion, viz. of Bleeding a Mangy into a Sound Dog." May 6, 1667: 451–452.

"An Account of More Tryals of Transfusion, Accompanied with Some Considerations Thereon, Chiefly in Reference to its Circumspect Practise on Man; Together with a Farther Vindication of This Invention from Usurpers." October 21, 1667: 517–525.

"An Account of the Experiment of Transfusion, Practised Upon a Man in London." December 9, 1667: 557–559.

"An Advertisement Concerning the Invention of the Transfusion of Bloud." July, August, September 1667: 489–490.

"Of the Antiquity of the Transfusion of Bloud from One Animal to Another." July 31, 1668, 731–732.

"Concerning a New Way of Curing Sundry Diseases by Transfusion of Blood, Written to Monsieur de Montmor, Counsellor to the French King, and Master of Requests." June 25, 1667: 489–504.

"Experiments Made at London Concerning the Liquor Sent out of France, Which Is There Famous for Staunching of the Blood of Arteries as Well as Veins." 1673, vol. 8: 6052–6059

"An Extract of a Letter of M. Denis . . . to M*** Touching the Transfusion of Blood, of April 2, 1667." 1666: 453.

"An Extract of a Letter, Written by J. Denis, Doctor of Physick, and Professor of Philosophy and the Mathematicks at Paris, Touching a Late Cure of an Inveterate Phrensy by the Transfusion of Blood." February 10, 1668: 617–623.

"Extract of a Letter, Written to the Publisher by M. Denys from Paris; Giving Notice of an Admirable Liquor, Instantly Stopping the Blood of Arteries Prickt or Cut, Without Any Suppuration, or Without Leaving Any Scar or Cicatrice." 1673, vol. 8: 6039.

"An Extract of a Printed Letter, Addressed to the Publisher, by M. Jean Denis. . . . Touching on the Differences Risen About the Transfusion of Bloud." 1668, vol. 3: 710–715.

"An Extract out of the Italian Giornale de Letterati, About Two Considerable Experiments of the Transfusion of the Blood." May 8, 1667: 840–842.

"A Letter, Written to the Publisher by the Learned and Experienced Dr. Timothy Clarck, One of His Majesties Physicians in Ordinary,

Concerning Some Anatomical Inventions and Observations, Particularly the Origin of the Injection into Veins, the Transfusion of Bloud, and the Parts of Generation." May 18, 1668: 672–682.

"The Method Observed in Transfusing the Bloud out of One Animal into Another." December 17, 1666: 353–358.

"A Relation of Some Trials of the Same Operation, Lately Made in France." December 9, 1667: 559–564.

"The Success of the Experiment of Transfusion the Bloud of One Animal into Another." November 19, 1666: 352.

"Trials Proposed by Mr. Boyle to Dr. Lower, to Be Made by Him, for the Improvement of Transfusing Bloud out of One Live Animal into Another." February 11, 1667: 385–388.

Pliny. *Natural History.* 10 vols. Edited and translated by Harris Rackman. Cambridge, MA: Harvard University Press, 1938–1963.

Poterie, Antoine de la. Letter on transfusion. 28 December 1667.

Prévost, J. L., and J. B. Dumas. "Examen du sang et de son action dans les divers phénomènes de la vie." *Annales chimiques* 18 (1821): 280.

Recueil de quelques nouvelles observations des effets très considérables de la transfusion du sang et de l'infusion des médicaments dans les veines. The Hague, 1668.

"Red Cross to Use Blood of Negroes." *New York Times,* January 29, 1942: 13.

Rush, Benjamin. "A Defence of Blood-Letting as a Remedy for Certain Diseases." *Medical Inquiries and Observations.* 5 vols. Philadelphia, 1794–1798. Vol. 4, 183–258.

Santinello, Bartholomaeo. *Confusio transfuionis sive confutatio operationis transfundentis sanguinem de individuo ad individuum.* Rome, 1668.

Sauval, Henri. *Histoire et recherches des antiquités de la ville de Paris.* 3 vols. 1724. Reprint, Geneva: Minkoff, 1973.

Scudder, John, Shivaji B. Bhonslay, Aaron Himmelstein, and John G. Gorman. "Sensitising Antigens as Factors in Blood Transfusions: The Complicating Factor of an Anti-Kidd (JK[a]) Antibody in a Patient with Myxoma of the Left Auricle Undergoing Open-Heart Surgery." Paper presented by Dr. Gorman at the Twelfth Annual Meeting of the American Association of Blood Banks, Chicago, November 6, 1959.

Sévigné, Marie de. *Lettres de Mme de Sévigné.* Paris: Firmin Didot Frères, 1843.

Sorbière, Samuel de. *Relation d'un voyage en Angleterre.* Paris, 1664.

———. *Relations, lettres et discours de M. de Sorbière.* Paris, 1666.

Sprat, Thomas. *The History of the Royal Society of London for the Improving of Natural Knowledge.* 3rd ed., London, 1722.

Stubbe, Henry. *The Plus Ultra reduced to a Non Plus: Or, A Specimen of Some Animadversions Upon the Plus Ultra of Mr. Glanvill, wherein Sundry Errors of Some Virtuosi are Discovered.* n.p., 1680.

Tallement des Réaux. *Historiettes.* Edited by Georges Mongredien. Paris: Librairie Garnier, 1932.

Taswell, William. *Autobiography and Anecdotes.* London, Camden Society, 1852.

Thornton, William. *Papers of William Thornton.* Edited by C. M. Harris. Vol 1. Charlottesville: University Press of Virginia, 1995.

"Use of Negro Blood for Blood Banks." *JAMA* 3 (1942): 307–308.

Vattier, Pierre. *Le Coeur détroné, discours de l'usage du foye où il est monstré que le coeur ne fait pas le sang, & qu'il n'est pas mesme une des principales parties de l'animal, prononcé dans une assemblée de physiciens chez Monsieur de Montmor.* Paris, 1650

Vincent, Thomas. *God's Terrible Voice in the City.* London, 1667.

Voltaire. *Le Siècle de Louis XIV.* 1751. Reprint, Paris: Garnier Frères, 1947.

W. M. *The Queen's Closet Opened.* London, 1655.

Willis, Thomas. *Cerebri anatome.* London, 1664.

———. *Two Discourses Concerning the Soul of Brutes, which is that of the Vital and Sensitive of Man.* Translated by S. Pordage. London, 1683.

Wood, Anthony. *The Life and Times of Anthony Wood, antiquary, of Oxford, 1632–1695, described by himself.* Edited by Andrew Clark and Llewelyn Powys. 5 vols. Oxford: Oxford Historical Society, 1891–1900.

Woolley, Hannah. *The Queen-Like Closet.* London, 1675.

Wren, Christopher, Jr. *Parentalia: Or Memoirs of the Family of the Wrens.* London, 1750.

SECONDARY SOURCES

Ackerknecht, Erwin H. *A Short History of Medicine.* Baltimore: Johns Hopkins University Press, 1982.

———. *Therapeutics: From the Primitives to the 20th Century*. New York: Hafner Press, 1973.

Adkins, G. Matthew. "The Montmor Discourse: Science and the Ideology of Stability in Old Régime France." *Journal of the Historical Society* 5 (2005): 1–28.

Allderidge, Patricia. "Bedlam: Fact or Fantasy?" In *The Anatomy of Madness: Essays in the History of Psychiatry*, edited by W. F. Bynum, R. Porter, and M. Sheperd. Vol. 2. 17–33. London: Tavistock, 1985.

Anderson, R. G. W., J. A. Bennett, and W. F. Ryan, eds. *Making Instruments Count: Essays on Historical Scientific Instruments Presented to Gerard L'Estrange Turner.* Aldershot, UK: Variorum, 1993.

Andrade, E. N. da C. "The Birth and Early Days of the *Philosophical Transactions.*" *Notes and Records of the Royal Society of London.* 20 (1965): 9–27.

Andrews, Richard Mowery. *Law, Magistracy, and Crime in Old Regime Paris, 1735–1789.* Vol. 1. Cambridge, UK: Cambridge University Press, 1994.

Andriesse, Cornelius Dirk. *Huygens: The Man Behind the Principle.* Cambridge, UK: Cambridge University Press, 2005.

Arikha, Noga. "Form and Function in the Early Enlightenment." *Perspectives on Science* 14 (2006): 153–188.

———. *Passions and Tempers: A History of the Humours.* New York: Ecco Press, 2007.

Astier, Régine. "Louis XIV, 'Premier Danseur.' " In *Sun King: The Ascendancy of French Culture During the Reign of Louis* XIV, edited by David Lee Rubin, 73–102. Washington, DC: Folger Books, 1992.

Augarde, Jean-Dominique. "La Fabrication des instruments scientifiques au XVIIIe siècle et la corporation des fondeurs." In *Studies in the History of Scientific Instruments: Papers Presented at the 7th Symposium of the Scientific Instruments Commission*, edited by Christine Blondel et al., 52–72. London: Rogers Turner Books, 1989.

Babington, Anthony. "Newgate in the Eighteenth Century." *History Today* 9 (1971): 650–657.

Balz, Albert G. A. *Cartesian Studies.* New York: Columbia University Press, 1951.

Baskett, Thomas F. "James Blundell: The First Transfusion of Human Blood." *Resuscitation* 52 (2002): 229–233.

Bates, Don G. "Harvey's Account of His 'Discovery.' " *Medical History* 36 (1992): 361–378.

Bell, Arthur E. *Christian Huygens and the Development of Science in the Seventeenth Century.* London: Edward Arnold, 1947.

Bell, David A. *Lawyers and Citizens: The Making of a Political Elite in the Old Regime.* Oxford: Oxford University Press, 1994.

Bell, Walter George. *The Great Fire of London in 1666.* London: John Lane, 1920.

Bennett, James A. *The Divided Circle: A History of Instruments for Astronomy, Navigation and Surveying.* Oxford: Phaidon/Christies, 1987.

Bigourdan, M.G. *Les Premières sociétés savantes de Paris au XVIIe siècle et les Origines de l'Académie des sciences.* Paris: Gauthier-Villars, 1918.

Birn, Raymond. *Le Journal des Savants sous l'Ancien Régime.* Paris: Editions Klincksieck, 1965.

Bluche, François, ed. *Dictionnaire du grand siècle.* Paris: Fayard, 1990.

———. *Louis XIV.* Translated by Mark Greengrass. New York: Blackwell, 1990.

Bonnicksen, Andrea L. *Chimeras, Hybrids, and Interspecies Research: Policy and Policy Making.* Georgetown: Georgetown University Press, 2009.

Brockliss, Laurence. "Medical Teaching at the University of Paris, 1660–1720." *Annals of Science* 35 (1978): 221–251.

———, and Colin Jones. *The Medical World of Early Modern France.* Oxford: Clarendon Press, 1997.

Brown, Harcourt. "Jean Denis and the Transfusion of Blood, Paris, 1667–1668." *Isis* 39 (1947): 15–29.

———. *Scientific Organizations in Seventeenth-Century France (1620–1680).* New York: Russell & Russell, 1934.

Browne, Christopher. *Getting the Message: The Story of the British Post Office.* Dover: Alan Sutton, 1993.

Brygoo, Edouard. "Les médecins de Montpellier et le Jardin du Roi à Paris." *Histoire et nature* 14 (1979): 3–29.

Burke, Peter. *The Fabrication of Louis XIV.* New Haven: Yale University Press, 1992.

Bylebyl, Jerome. "Boyle and Harvey on the Valves in the Veins." *Bulletin of the History of Medicine* 56 (1982): 351–367.

Bynum, William F. "The Anatomical Method, Natural Theology, and the Functions of the Brain." *Isis* 64 (1973): 445–468.

Chardans, Jean-Louis. *Le Châtelet: de la prison au théâtre.* Paris: Pygmalion, 1980.

Chassagne, Annie. *La Bibliothèque de l'Académie Royale des Sciences au XVIIIe siècle.* Paris: Éditions du comité des travaux historiques et scientifiques, 2007.

Chauvois, L. "Le Docteur Pierre Vattier (1623–1670)." *La Presse médicale* 37 (1955): 1887–1888.

Clarke, Edwin, and Kenneth Dewhurst. *An Illustrated History of Brain Function.* Berkeley: University of California Press, 1972.

Clayton, Jay. "Victorian Chimeras, or, What Literature Can Contribute to Genetics Policy Today." *New Literary History* 3 (2007): 569–591.

Cockayne, Emily. *Hubbub: Filth, Noise, and Stench in England.* New Haven: Yale University Press, 2007.

Collas, George. *Jean Chapelain, 1595–1674.* Paris: Perrin et Cie, 1911.

Collins, James. *The State in Early Modern France.* Cambridge, UK: Cambridge University Press, 1995.

Daumas, Maurice. *Scientific Instruments of the 17th and 18th Centuries and Their Makers.* Translated by Mary Holbrook. London: Portman Books, 1972.

Debus, Allen G. *The French Paracelsians: The Chemical Challenge to Medical and Scientific Tradition in Early Modern France.* Cambridge, UK: Cambridge University Press, 1991.

DeJean, Joan. *The Essence of Style: How the French Invented High Fashion, Fine Food, Chic Cafés, Style, Sophistication, and Glamour.* New York: Free Press, 2005.

Delorme, Suzanne. "Un Cartésien ami: Henri-Louis de la Martinière." *Revue d'histoire des sciences* 27 (1974): 68–72.

Dessert, Daniel. *Fouquet.* Paris: Fayard, 1987.

Diamond, Louis K. "A History of Blood Transfusion." In *Blood, Pure and Eloquent: A Story of Discovery, of People, and of Ideas,* edited by Maxwell M. Wintrob, 659–688. New York: McGraw-Hill, 1980.

Dolan, Frances E. "Ashes and the 'Archive': The London Fire of 1666, Partisanship and Proof." *Journal of Medieval and Early Modern Studies* 31 (2001): 379–408.

Dulieu, Louis. *La Médicine à Montpellier.* Avignon: Presses universelles, 1975.

Dunlop, Ian. *Louis XIV.* London: Chatto & Windus, 1999.

Eamon, William. *Science and the Secrets of Nature.* Princeton: Princeton University Press, 1994.

Ellis, R. W. B. "Blood Transfusion at the Front." *Proceedings of the Royal Society of Medicine* 31 (1938): 684–686.

Elmer, Peter, ed. *The Healing Arts: Health, Disease and Society in Europe, 1500–1800.* Manchester, UK: Open University, 2004.

Fagan, Brian M. *The Little Ice Age: How Climate Made History, 1300–1850.* New York: Basic Books, 2001.

Farr, A. D. "The First Human Blood Transfusion." *Medical History* 24 (1980): 143–162.

Fauré-Fermiet, E. "Les Origines de l'Académie des Sciences de Paris." *Notes and Records of the Royal Society of London* 21 (1966): 20–31.

Faustin, Foiret. "L'Hôtel de Montmor." *La Cité, bulletin trimestriel de la Société historique et archéologique du IVe arrondissement de Paris* 13 (1914): 309–339.

Frank, Robert G. "Viewing the Body: Reframing Man and Disease in Commonwealth and Restoration England." In *The Restoration Mind,* edited by W. Gerald Marshall, 65–110. Newark: University of Delaware Press, 1997.

———. *Harvey and the Oxford Physiologists: A Study of Scientific Ideas.* Berkeley: University of California Press, 1980.

Garber, Daniel. "Descartes, Mechanics, and the Mechanical Philosophy." *Midwest Studies in Philosophy* 26 (2002): 185–204.

———. "Soul and Mind: Life and Thought in the Seventeenth Century." In Daniel Garber, Roger Ariew, and Michael Ayers, eds., *The Cambridge History of Seventeenth-Century Philosophy.* New York and Cambridge, UK: Cambridge University Press, 1998: 759–795.

Garnot, B. *La Population française aux XVIe, XVIIe, XVIIIe siècles.* Paris: Editions Orphys, 1988.

George, Albert Joseph. "A Seventeenth-Century Amateur of Science: Jean Chapelain." *Annals of Science* 3 (1938): 217–236.

Gibson, William Carleton. "The Bio-Medical Pursuits of Christopher Wren." *Medical History* 14 (1990): 331–341.

Gottlieb, A. M. "History of the First Blood Transfusion but a Fable Agreed

Upon: The Transfusion of Blood to a Pope." *Transfusion Medicine Reviews* 5 (1991): 228–235.

Greely, Henry T., Mildred K. Cho, Linda F. Hogle, and Debra M. Satz. "Thinking about the Human Neuron Mouse." *American Journal of Bioethics* 7, no. 5 (2007): 27–40.

Greenberger, Gerald A. "Lawyers Confront Centralized Government: Political Thought of Lawyers During the Reign of Louis XIV." *American Journal of Legal History* 23 (1979): 144–181.

Greene, Mark, Kathryn Schill, Shoji Takahashi, Alison Bateman-House, Tom Beauchamp, Hilary Bok, Dorothy Cheney, Joseph Coyle, Terrence Deacon, Daniel Dennett, Peter Donovan, Owen Flanagan, Steven Goldman, Henry Greely, Lee Martin, Early Miller, Dawn Mueller, Andrew Mueller, Andrew Siegel, Davor Solter, John Gearhart, Guy McKhann, and Ruth Faden. "Moral Issues of Human-Non-Human Primate Neural Grafting." *Science* 309 (2005): 385–386.

Gribbin, John. *The Fellowship: Gilbert, Bacon, Harvey, Wren, Newton, and the Story of a Scientific Revolution.* New York: Overlook Press, 2007.

Grifols, Joan R. "The Contribution of Dr. Duran-Jordá to the Advancement and Development of European Blood Transfusion." *ISBT Science Series* 2 (2007): 134–138.

Guerrini, Anita. "The Ethics of Animal Experimentation in Seventeenth-Century England." *Journal of the History of Ideas* 50 (1989): 391–407.

Gunson, Harold H., and Helen Dodsworth. "Fifty Years of Blood Transfusion." *Transfusion Medicine* 6 (1996): 1–88.

Gunther, R. T. *Early Science in Oxford.* Vols. 77–78. Oxford: Oxford Historical Society, 1923.

Hahn, Roger. *The Anatomy of a Scientific Institution: The Paris Academy of Sciences, 1666–1803.* Berkeley: University of California Press, 1971.

———. "Changing Patterns for the Support of Scientists from Louis XIV to Napoleon." *History and Technology* 4 (1987): 401–411.

———. "Louis XIV and Science Policy." In *Sun King: The Ascendancy of French Culture During the Reign of Louis XIV,* edited by David Lee Rubin, 195–206. Washington, DC: Folger Books, 1992,

Hall, Marie Boas. *Henry Oldenburg: Shaping the Royal Society.* Oxford: Oxford University Press, 2002.

———. "Oldenburg and the Art of Scientific Communication." *British Journal for the History of Science* 2 (1965): 277–290.

———. "The Royal Society's Role in the Diffusion of Information in the Seventeenth Century." *Notes and Records of the Royal Society of London* 29 (1975): 173–192.

Hall, Rupert A. "English Medicine in the Royal Society's Correspondence: 1660–1677." *Medical History* 15 (1971): 111–125.

———, and Marie Boas Hall. "The First Human Blood Transfusion: Priority Disputes." *Medical History* 24 (1980): 461–465.

Hallay, André. *Les Perraults.* Paris: Perrin et cie, 1926.

Hamscher, Albert N. *The Parlement of Paris after the Fronde, 1653–1673.* Pittsburgh: University of Pittsburgh Press, 1976.

Harris, Henry. *The Birth of the Cell.* New Haven: Yale University Press, 2000.

Hay, Douglas, Peter Linebaugh, John C. Rule, F. P. Thompson, and Cal Winston, eds. *Albion's Fatal Tree: Crime and Society in Eighteenth Century England.* New York: Pantheon, 1975.

Hazard, Jean. "Claude Perrault, architecte célèbre, médecin méconnu, chercheur infatigable." *Histoire des sciences médicales* 41 (2007): 399–406.

Herrmann, Wolfgang. *The Theory of Claude Perrault.* London: A. Zwemmer, 1973.

Hillairet, Jacques. *Dictionnaire historique des rues de Paris.* 2 vols. Paris: Éditions de Minuit, 1985.

Hillyer, Christopher D., Beth H. Shaz, James C. Zimring, and Thomas C. Abshire, eds. *Transfusion Medicine and Hemostasis: Clinical and Laboratory Aspects.* Amsterdam: Elsevier, 2009.

Hollis, Leo. *London Rising: The Men Who Made Modern London.* New York: Walker, 2008.

Howard, Nicole. "Rings and Anagrams: Huygens' System of Saturn." *Papers of the Bibliographical Society of America* 98 (2004): 477–510.

Hunt, Margaret. "Hawkers, Bawlers, and Mercuries: Women and the London Press in the Early Enlightenment." *Women & History* 9 (1984): 41–68.

Hunter, Michael. "Alchemy, Magic and Moralism in the Thought of Robert Boyle." *British Journal for the History of Science* 23 (1990): 387–410.

———. "Promoting the New Science: Henry Oldenburg and the Early Royal Society." *History of Science* 25 (1988): 165–180.

Hussey, Andrew. *Paris: The Secret History*. New York: Bloomsbury, 2006.

Hutchinson, Harold F. *Sir Christopher Wren*. New York: Stein & Day, 1976.

Jardine, Lisa. *The Curious Life of Robert Hooke: The Man Who Measured London*. New York: HarperCollins, 2004.

———. "Dr. Wilkins' Boy Wonders." *Notes and Records of the Royal Society of London* 58, no. 1 (2004): 107–129.

———. *Ingenious Pursuits: Building the Scientific Revolution*. New York: Doubleday, 1999.

———. *On a Grander Scale: The Outstanding Life of Sir Christopher Wren*. New York: HarperCollins, 2002.

Jenner, Mark. "The Politics of London Air: John Evelyn's *Fumifugium* and the Restoration." *Historical Journal* 38 (1995): 535–551.

Johns, Adrian. "Miscellaneous Methods: Authors, Societies, and Journals in Early Modern England." *British Journal for the History of Science* 2 (2000): 159–186.

Jones, Colin. "Montpellier Medical Students and the Medicalisation of 18th Century France." In *Problems and Methods in the History of Medicine*, edited by Roy Porter and Andrew Wear, 57–80. London: Croom Helm, 1987.

Kalof, Linda. *Looking at Animals in Human History*. London: Reaktion, 2007.

Karpowitz, Phillip, Cynthia B. Cohen, and Derek van der Kooy. "Developing Human-Nonhuman Chimeras in Human Stem Cell Research: Ethical Issues and Boundaries." *Kennedy Institute of Ethics Journal* (2005) 15: 107–134.

Keele, Kenneth D. *William Harvey: The Man, the Physician, and the Scientist*. London: Thomas Nelson, 1965.

Kerviler, René. "Henri-Louis Habert de Montmor." *Bibliophile français* (1872): 198–208.

Kettering, Sharon. *French Society, 1589–1715*. Harlow, UK: Longman, 2001.

———. *Patrons, Brokers, and Clients in Seventeenth-Century France*. New York: Oxford University Press, 1986.

Keynes, Geoffrey. *Blood Transfusion*. Baltimore: Williams & Wilkins, 1949.

———. *The Life of William Harvey*. Oxford: Clarendon Press, 1966.

Kleinman, Ruth. *Anne of Austria: Queen of France*. Columbus: Ohio State University Press, 1985.

Knight, Harriet. "Robert Boyle's *Memoirs for the Natural History of Human Blood* (1684): Print, Manuscript, and the Impact of Baconianism in Seventeenth-Century Medical Science." *Medical History* 51 (2007): 145–164.

Knoeff, Rina. "The Reins of the Soul: The Centrality of the Intercostal Nerves to the Neurology of Thomas Willis and to Samuel Parker's Theology." *Journal of the History of Medicine and Allied Sciences* 59 (2004): 413–440.

Koslofsky, Craig. "Court Culture and Street Lighting in Seventeenth-Century Europe." *Journal of Urban History* 28 (2002): 743–768.

Kuriyama, Shigehisa. "Interpreting the History of Bloodletting." *Journal of the History of Medicine and Allied Sciences* 50 (1995): 11–46.

Lawrence, Christopher, and Steven Shapin. *Science Incarnate: Historical Embodiments of Nature Knowledge*. Chicago: University of Chicago Press, 1998.

Lederer, Susan E. *Flesh and Blood: Organ Transplantation and Blood Transfusion in Twentieth-Century America*. Oxford: Oxford University Press, 2008.

———. *Subjected to Science: Human Experimentation in America Before the Second World War*. Baltimore: Johns Hopkins University Press, 1995.

Lennon, Thomas M. *The Battle of the Gods and Giants: The Legacies of Descartes and Gassendi, 1655–1715*. Princeton: Princeton University Press, 1993.

Leshner, Alan I. "Where Science Meets Society." *Science* 307 (2005): 815.

Lévy-Valensi, J. *La Médicine et les médecins au XVIIe siècle*. Paris: J.-B. Baillière et fils, 1933.

Lindeboom, G. A. "The Story of a Blood Transfusion to a Pope." *Journal of the History of Medicine* (1954), October: 455–459.

Linebaugh, Peter. "The Tyburn Riot Against the Surgeons." In *Albion's Fatal Tree: Crime and Society in Eighteen Century England*, edited by Douglas Hay, Peter Linebaugh, John C. Rule, F. P. Thompson, and Cal Winston. 65–117. London: Allen Lane, 1975.

Loux, Françoise. *Pierre-Martin de la Martinière, un médecin au xviie siècle*. Paris: Editions Imago, 1988.

Lux, David. "Colbert's Plan for the Grande Académie." *Seventeenth-Century French Studies* 12 (1990): 177–188.

Maehle, Andreas Holger, and Ulrich Tröhler. "Animal Experimentation from Antiquity to the End of the Eighteenth Century: Attitudes and Arguments." In *Vivisection in Historical Perspective*, edited by Nicolaas A. Rupke, 14–47. London: Taylor & Francis, 1987.

Maindron, Ernest. *L'Académie des Sciences*. Paris: Félix Alcan, 1888.

Maluf, N. S. R. "History of Blood Transfusion." *Journal of the History of Medicine* (1954): 59–107.

Marshall, Alan. *Intelligence and Espionage in the Reign of Charles II, 1660–1685*. Cambridge, UK: Cambridge University Press, 1994.

Martensen, Robert Lawrence. *The Brain Takes Shape: An Early History*. Oxford: Oxford University Press, 2004.

Martin, D. C. "Former Homes of the Royal Society." *Notes and Records of the Royal Society* 22 (1967): 12–19.

Mattern, Susan P. *Galen and the Rhetoric of Healing*. Baltimore: Johns Hopkins University Press, 2008.

Mayor, Adrienne. *The Poison King: The Life and Legend of Mithradates*. Princeton: Princeton University Press, 2010.

McDonald, Michael. *Mystical Bedlam: Madness, Anxiety, and Healing in Seventeenth-Century England*. Cambridge, UK: Cambridge University Press, 1981.

McKie, Douglas. "The Arrest and Imprisonment of Henry Oldenburg." *Notes and Records of the Royal Society of London* 6 (1948): 28–47.

McMullen, Emerson Thomas. "Anatomy of a Physiological Discovery: William Harvey and the Circulation of the Blood." *Journal of the Royal Society of Medicine* 88 (1995): 491–498.

Meynell, Guy. "The Académie des Sciences at the Rue Vivienne, 1666–1699." *Archives Internationales d'Histoire des Sciences* 44 (1994): 22–37.

Montclos, Jean Marie Perouse de. *Vaux Le Vicomte*. London: Editions Scala, 1997.

Moore, Pete. *Blood and Justice: The Seventeenth-Century Doctor Who Made Blood Transfusion History*. New York: John Wiley, 2003.

Moote, A. Lloyd, and Dorothy C. Moote. *The Great Plague: The Story of London's Most Deadly Year*. Baltimore: Johns Hopkins University Press, 2004.

Morabia, Alfredo. "P. C. A. Louis and the Birth of Clinical Epidemiology." *Journal of Clinical Epidemiology* 49 (1996): 1327–1333.

Moran, Bruce T. *Distilling Knowledge: Alchemy, Chemistry, and the Scientific Revolution*. Cambridge, MA: Harvard University Press, 2005.

Morens, David. "Death of a President." *New England Journal of Medicine* 341 (1999): 1845–1850.

Morize, André. "Samuel Sorbière (1610–1670)." *Zeitschrift für französiche Sprache und Literatur* 3 (1908): 239–257.

Mousnier, Roland. *The Institutions of France Under the Absolute Monarchy, 1598–1789*. Chicago: University of Chicago Press, 1984.

Newman, William R. "From Alchemy to 'Chymistry.' " *Early Modern Science*, edited by Katharine Park and Lorraine Daston, 497–517. Cambridge, UK: Cambridge University Press, 2006.

———, and Lawrence M. Principe. "Alchemy vs. Chemistry: The Etymological Origins of a Historiographic Mistake." *Early Science and Medicine* 3 (1998): 32–65.

Nicolson, Marjorie Hope. *Nicolson. Pepys' Diary and the New Science*. Charlottesville: University of Virginia Press, 1965.

Nomblot, Jean. *Pierre Martin de la Martinière (1634–1676): Médecin empirique du XVIIe siècle*. Paris: Librairie du Vieux-Colombier, 1932.

Numbers, Ronald L. *Galileo Goes to Jail and Other Myths About Science and Religion*. Cambridge, MA: Harvard University Press, 2009.

Ollard, Richard. *Pepys: A Biography*. New York: Holt, Rinehart & Winston, 1975.

Park, Katharine. "Myth 5: That the Medieval Church Prohibited Human Dissection." In *Galileo Goes to Jail and Other Myths About Science and Religion*, edited by Ronald L. Numbers, 43–49. Cambridge, MA: Harvard University Press, 2009.

———. "Psychology: The Organic Soul." In *The Cambridge History of Renaissance Philosophy*, edited by Charles Schmitt, Quentin Skinner, Eckhard Kessler, and Jill Kraye, 476–484. Cambridge, UK: Cambridge University Press, 1988.

Pelis, Kim. "Blood Clots: The Nineteenth-Century Debate Over the Substance and Means of Transfusion." *Annals of Science* 54 (1997): 331–360.

———. "Taking Credit: The Canadian Army Medical Corps and the British

Conversation to Blood Transfusion in WWI." *Journal of the History of Medicine* 56 (2001): 238–277.

Perkins, Wendy. "The Uses of Science: The Montmor Academy, Samuel Sorbière, and Francis Bacon." *Seventeenth Century* 7 (1985): 155–162.

Peters, Edward M. "Prison Before the Prison: The Ancient and Medieval Worlds." In *The Oxford History of the Prison: The Practice of Punishment in Western Society*, edited by Norval Morris and David J. Rothman, 3–47. Oxford: Oxford University Press, 1995.

Peumery, J. J. "Conversations médico-scientifiques de l'Académie de l'Abbé Bourdelot." *Histoire des Sciences Médicales* 12 (1978): 127–135.

———. *Jean-Baptiste Denis et la recherche scientifique au XVIIe siècle.* Paris: L'Expansion scientifique française, 1970.

———. *Les Origines de la transfusion sanguine.* Amsterdam: B. M. Israël, 1975.

Pickstone, John V. "Globules and Coagula: Concepts of Tissue Formation in the Early Nineteenth Century." *Journal of the History of Medicine and Allied Sciences* 4 (1973): 336–356.

Picon, Antoine. *Claude Perrault, ou la curiosité d'un classique.* Paris: Picard, 1988.

Porter, Roy. *Blood and Guts: A Short History of Medicine.* New York: W. W. Norton, 2004.

———. *Flesh in the Age of Reason: The Modern Foundations of Body and Soul.* New York: W. W. Norton, 2005.

———. *The Greatest Benefit to Mankind: A Medical History of Humanity.* New York: W. W. Norton, 1997.

———. *Mind-forg'd Manacles: A History of Madness in England from the Restoration to the Regency.* Cambridge, MA: Harvard University Press, 1987.

Porter, Stephen. *The Great Fire of London.* London: Alan Sutton, 1997.

Principe, Lawrence M. *The Aspiring Adept: Robert Boyle and His Alchemical Quest.* Princeton: Princeton University Press, 1998.

———. "Boyle's Alchemical Pursuits." In *Robert Boyle Reconsidered*, edited by Michael Hunter, 91–105. Cambridge, UK: Cambridge University Press, 1994.

———. "Newly Discovered Boyle Documents in the Royal Society Archive: Alchemical Tracts and His Student Notebook." *Notes and Records of the Royal Society of London* 49 (1995): 57–70.

Ramey, Lynn. "Monstrous Alterity in Early Modern Travel Accounts: Lessons from the Ambiguous Medieval Discourses on Humanness." *Esprit Créateur* 48 (2008): 81–95.

Ranum, Orest. *Paris in the Age of Absolutism*. 2nd ed. University Park: Pennsylvania State University Press, 2002.

Reddaway, T. F. *Rebuilding London After the Great Fire*. London: Jonathan Cape, 1940.

Riesman, D. "Bourdelot, a Physician of Queen Christina of Sweden." *Annals of Transfusion* 9 (1937): 191.

Riley, Philip F. *A Lust for Virtue: Louis XIV's Attack on Sin in Seventeenth-Century France*. Westport, CT: Greenwood Press, 2001.

Risse, Geunther B. "The Renaissance of Bloodletting: A Chapter in Modern Therapeutics." *Journal of the History of Medicine and Allied Sciences* 34 (1979): 3–22

Rivington, Charles A. "Early Printers to the Royal Society, 1663–1708." *Notes and Records of the Royal Society of London* 39 (1984): 1–27.

Robb-Smith, A. H. T. "Unravelling the Functions of the Blood." *Medical History* 6 (1962): 1–21.

Roger, Jacques. *The Life Sciences in Eighteenth-Century French Thought*. Edited by Keith R. Benson. Translated by Robert Ellrich. 1963. Reprint, Stanford: Stanford University Press, 1997.

Rullière, R. "Le 'Tractatus de corde item de motu et colore sanguinis' de Richard Lower (1669)." *Histoire des Sciences Médicales* 8 (1974): 85–98.

Saint-Germain, Jacques. *La Reynie et la police au Grand Siècle*. Paris: Hachette, 1962.

Sarasohn, Lisa T. "Who Was Then the Gentleman?: Samuel Sorbière, Thomas Hobbes, and the Royal Society." *History of Science* 42 (2004): 211–232.

Schechner, Sara Genuth. "The Material Culture of Astronomy in Daily Life: Sundials, Science, and Social Change." *Journal of the History of Astronomy* 32 (2001): 189–222.

Schiller, Joseph. "La Transfusion sanguine et les débuts de l'Académie des Sciences." *Clio Medica* 1 (1965): 33–40.

Schiller, Netty. *Comets, Popular Culture, and the Birth of Modern Cosmology*. Princeton: Princeton University Press, 1997.

———. "L'Iconographie de Claude Perrault (1613–1688)." *91e Congres des Sociétés Savantes* 1 (1966): 215–234.

Schmidt, Paul J. "Transfuse George Washington!" *Transfusion* 42 (2002): 275–277.

Schneider, Gary T. *The Culture of Epistolarity: Vernacular Letters and Letter-Writing in Early Modern England, 1500–1700*. Newark: University of Delaware Press, 2005.

Schneider, William H. "Blood Transfusion Between the Wars." *Journal of the History of Medicine* 58 (2003): 187–224.

———. "Blood Transfusion in Peace and War, 1900–1918." *Social History of Medicine* 10 (1997): 105–126.

———. "The History of Research on Blood Group Genetics: Initial Discovery and Diffusion." *History and Philosophy of the Life Sciences* 18 (1996): 277–303.

Senior, Matthew. "The Ménagerie and the Labyrinthe: Animals at Versailles, 1662–1672." In *Renaissance Beasts: Of Animals, Humans, and Other Wonderful Creatures*, edited by Erica Fudge, 208–232. Urbana: University of Illinois Press, 2004.

Shapin, Steven. *The Scientific Revolution*. Chicago: University of Chicago Press, 1996.

———, and Simon Schaffer. *Leviathan and the Air-Pump*. Princeton: Princeton University Press, 1985.

Snow, Stephanie J. *Blessed Days of Anaesthesia: How Anaesthetics Changed the World*. Oxford: Oxford University Press, 2008.

Soll, Jacob. *The Information Master: Jean-Baptiste Colbert's Secret State Intelligence System*. Ann Arbor: University of Michigan Press, 2009.

Starr, Douglas. *Blood: An Epic History of Medicine and Commerce*. New York: Alfred A. Knopf, 1998.

Stroup, Alice. *A Company of Scientists: Botany, Patronage, and Community at the Seventeenth-Century Parisian Royal Academy of Sciences*. Berkeley: University of California Press, 1990.

———. "Louis XIV as Patron of the Parisian Academy of Sciences." In *Sun King: The Ascendancy of French Culture During the Reign of Louis XIV*, edited by David Lee Rubin, 221–240. Washington, DC: Folger Books, 1992.

Sturdy, David. *Science and Social Status: The Members of the Académie des Sciences, 1666–1750.* Rochester, NY: Boydell Press, 1985.

Taton, René. "Huygens et l'Académie royale des sciences." In *Huygens et la France*, edited by René Taton, 57–68. Paris: Vrin, 1981.

———. *Les Origines de l'Académie Royale des Sciences.* Conférence donnée au Palais de la Découverte, Paris, 15 mai 1965.

Thomson, Anne. *Bodies of Thought: Science, Religion and the Soul in the Early Enlightenment.* Oxford: Oxford University Press, 2008.

———. "Guillaume Lamy et l'âme matérielle." *Le Materialisme des Lumières.* Paris: Presses Universitaires de France, 1992, 63–71.

Thrower, Norman J. W. "Samuel Pepys FRS (1633–1703) and the Royal Society." *Notes and Records of the Royal Society of London* 57 (2003): 3–13.

Tinniswood, Adrian. *By Permission of Heaven: The True Story of the Great Fire of London.* New York: Riverhead Books, 2003.

Trout, Andrew. *Jean-Baptiste Colbert.* Boston: Twayne, 1978.

Turner, Anthony. "An Interrupted Story: French Translations from *Philosophical Transactions* in the Seventeenth and Eighteenth Centuries." *Notes and Records of the Royal Society of London* 62 (2008): 341–354.

———. *Early Scientific Instruments: Europe 1400–1800.* London: Sotheby's, 1987.

Van Helden, Albert. "Saturn and His Anses." *Journal for the History of Astronomy* 5 (1974): 105–121.

———. "The Telescope in the Seventeenth Century." *Isis* 65 (1974): 38–58.

Vila, Anne C. *Enlightenment and Pathology: Sensibility in the Literature and Medicine of Eighteenth-Century France.* Baltimore: Johns Hopkins University Press, 1998.

Wailoo, Keith. *Drawing Blood: Technology and Disease Identity in Twentieth-Century America.* Baltimore: Johns Hopkins University Press, 1997.

Walters, Barrie. "The *Journal des Savants* and the Dissemination of News of English Scientific Activity." *Studies on Voltaire and the Eighteenth-Century* 314 (1993): 133–166.

Walton, Michael. "The First Blood Transfusion: French or English?" *Medical History* 18 (1974): 360–364.

Walton, Michael T., and Phyllis J. Walton. "Witches, Jews, and Spagyrists:

Blood Remedies and Blood Transfusion in the Sixteenth Century." *Cauda Pavonis* 15 (1996): 12–15.

Watkins, W. M. "The ABO Blood Group System: Historical Background." *Transfusion Medicine* 11 (2001): 243–265.

Webster, Charles. "The Origins of Blood Transfusion: A Reassessment." *Medical History* 15 (1971): 387–392.

Weld, Charles Richard. *History of the Royal Society*. London: John W. Parker, 1848.

Westfall, Richard S. "Henri-Louis Habert de Montmor." *The Galileo Project*. http://galileo.rice.edu/Catalog/NewFiles/montmor.html (accessed June 1, 2010).

Whitteridge, Gweneth. *William Harvey and the Circulation of the Blood*. London: Macdonald, 1971.

Wiesner, Merry E. *The Marvelous Hairy Girls: The Gonzales Sisters and Their Worlds*. New Haven: Yale University Press, 2009.

Wright, John P. "The Embodied Soul in Seventeenth-Century Medicine." *Canadian Bulletin on Medicine* 8 (1991): 21–42.

Yeomans, Donald K. *Comets: A Chronological History of Observation, Science, Myth, and Folklore*. New York: John Wiley, 1991.

Zeller, G. "French Diplomacy and Foreign Policy in Their European Setting." In *The Ascendancy of France, 1648–1688*, vol. 5, *New Cambridge Modern History*, edited by F. L. Carsen, 198–221. Cambridge, UK: Cambridge University Press, 1961.

Zimmer, Carl. *The Soul Made Flesh: The Discovery of the Brain and How It Changed the World*. New York: Free Press, 2004.

Illustration Credits

Figure 1: Courtesy of the National Library of Medicine.
Figure 2: Courtesy of the National Library of Medicine.
Figure 3: Courtesy of the National Library of Medicine.
Figure 4: Courtesy of the National Library of Medicine.
Figure 5: Vanderbilt University, Eskind Biomedical Library Special Collections.
Figure 6: Vanderbilt University, Eskind Biomedical Library Special Collections.
Figure 7: Courtesy of the National Library of Medicine.
Figure 8: Courtesy of the National Library of Medicine.
Figure 9: University of Wisconsin–Madison, Special Collections.
Figure 10: Courtesy of the National Library of Medicine.
Figure 11: Wellcome Library, London.
Figure 12: © Académie des sciences—Institut de France.
Figure 13: © Académie des sciences—Institut de France.
Figure 14: University of Wisconsin–Madison, Special Collections.
Figure 15: Vanderbilt University, Jean and Alexander Heard Library.
Figure 16: Vanderbilt University, Jean and Alexander Heard Library.

Figure 17: Wellcome Library, London.
Figure 18: Courtesy of the National Library of Medicine.
Figure 19: Courtesy of the National Library of Medicine.
Figure 20: Courtesy of the National Library of Medicine.
Figure 21: Public domain.
Figure 22: Wellcome Library, London.

Index

Page numbers in *italics* refer to illustrations.